PROCESS PLANNING

PROCESS PLANNING

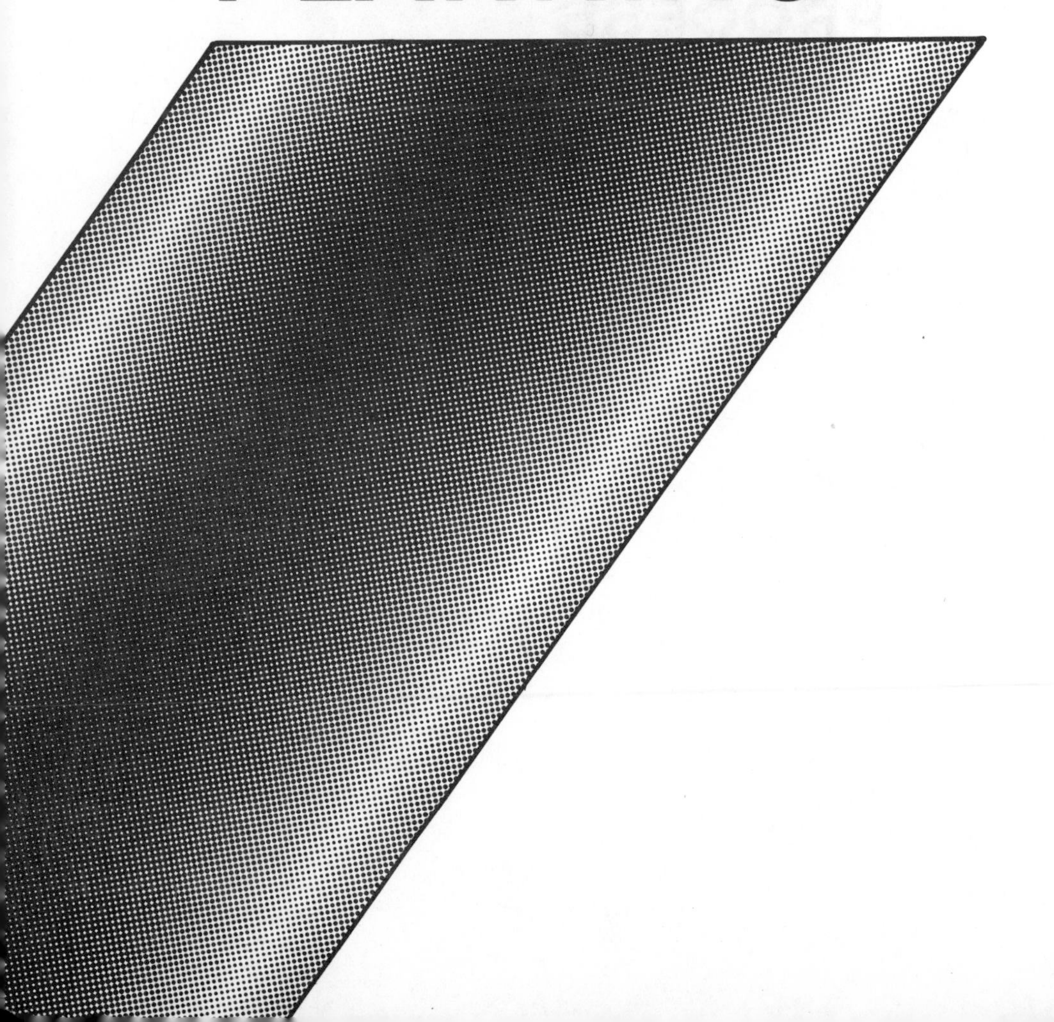

Mark A. Curtis

Ferris State University
Big Rapids, Michigan

John Wiley & Sons

New York • Chichester • Brisbane • Toronto • Singapore

Library of Congress Cataloging-in-Publication Data

Curtis, Mark A., 1951–
Process planning.

Bibliography: p.
1. Production engineering. I. Title.
TS176.C87 1988 658.5 87-34605
ISBN 0-471-83254-5

Printed in the United States of America

10 9 8 7 6 5 4 3 2 1

To Aaron and Leah

Preface

Today, talk of world competition, quality, and "high-tech" manufacturing techniques seems to be everywhere; American manufacturing industries are scrambling to turn this talk into real and productive action. However, industry knows that real changes to the face of modern manufacturing will come only with the help of many new and highly trained manufacturing professionals.

Many of the colleges and universities in the United States have responded to industry's cry for help with the creation of a variety of undergraduate and graduate programs in manufacturing engineering and technology. Central to these programs is the subject of process planning.

Therefore, as a course, Process Planning becomes extremely important. Yet, I believe that both teachers and students find it a very difficult subject to get a handle on. My intent in writing this book is to present the subject of process planning in a straightforward and organized manner.

Chapter 1 introduces the subject and its importance as related to overall manufacturing productivity. Chapters 2, 3, and 4 show how a process plan is actually made. Chapter 5 is on process documentation. Tolerance charting is covered in Chapter 6, and computer-aided process planning is addressed in Chapter 7. Chapter 8 discusses project management. Finally, the appendixes provide a wealth of reference information on materials, processes, and machines, each in a series of easy-to-read charts.

This book can be used as a comprehensive reference tool and text in a variety of engineering and technology programs where the subject of process planning is addressed. The book may also be used as a supplemental text in a wide range of manufacturing processes and cost-estimating courses.

In closing, I wish to acknowledge the support and encouragement given to me by my wife, family, colleagues, students, reviewers, and editor, Paul Berk, of John Wiley & Sons, Inc. A special thanks must also go to Robert

Duchesneau, a Ferris State College Manufacturing Engineering Technology Graduate and Outstanding Student, for his Computer-Aided Design work in preparing the appendixes. The author and publisher also acknowledge and thank the reviewers of the text for their comments and suggestions. The reviewers were James M. Fields, East Tennessee State University; Gary G. Hansen, Oklahoma State University; Charles R. Harrell, Brigham Young University; Harvey L. Hoy, Milwaukee School of Engineering; Wayne Lundberg, Rohr Industries, Inc.; Donald J. McAleece, Indiana University-Purdue University; Casimir Rakowski, SUNY Agricultural and Technical College; A Brent Strong, Brigham Young University; and John Steeves, Wentworth Institute.

Mark A. Curtis
Associate Professor
Manufacturing Engineering Technology
Ferris State University

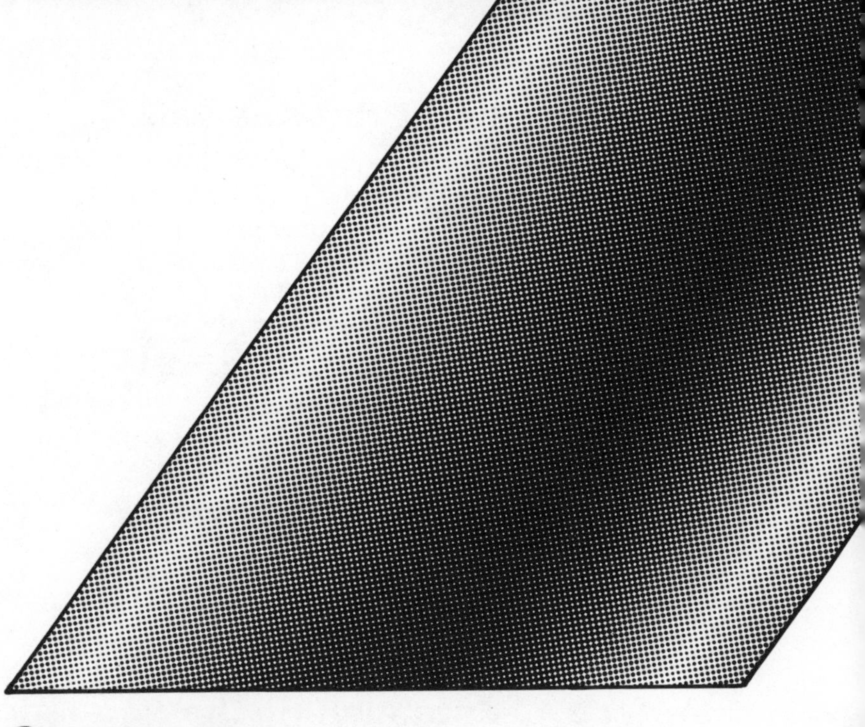

Contents

CHAPTER 1

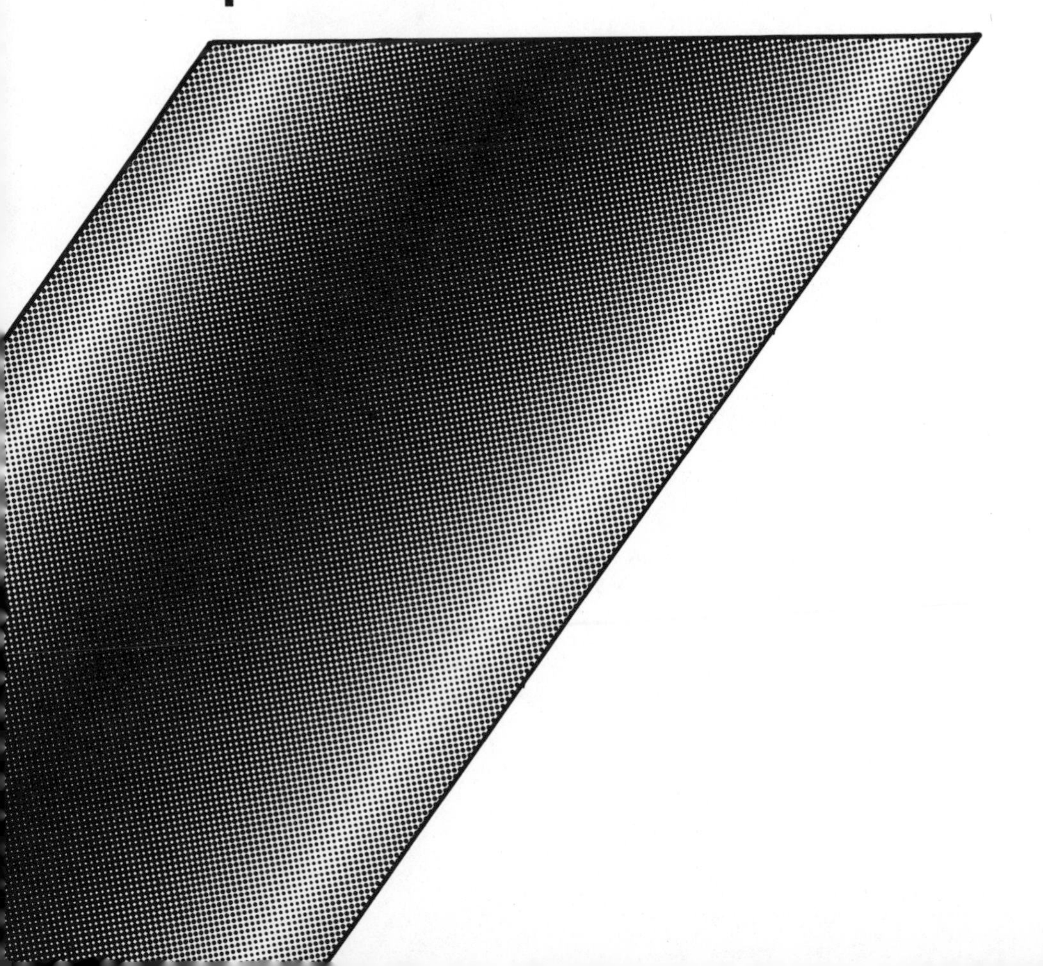

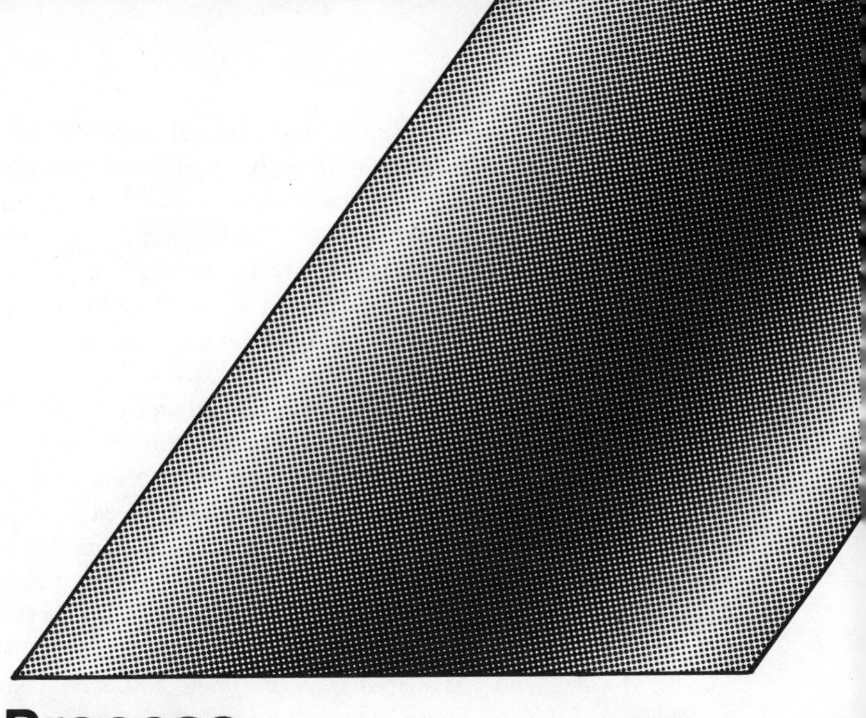

Process Planning

1 ▪ 1 INTRODUCTION

Process planning is the devising of a particular method of manufacturing, which generally involves a number of steps or operations. The process plan is a kind of road map to be followed in transforming raw materials into finished products.

The process-planning effort typically begins with an analysis of engineering drawings. These are detail drawings of the product under consideration for manufacture. The part's design specifications—including geometric features, dimensional sizes, tolerances, and material call-outs—must be evaluated in light of the production volume; that is, the total number of parts to be produced. The process plan that evolves from this preliminary review is generally considered a *cost estimate*. This estimate of the manufacturing costs is used by management to determine profitability and, in turn, the economic feasibility of producing the product being looked at. However, in the early stages, the sequence of manufacturing operations, including the machines and tooling required, are only tentative.

If the cost estimate, the result of the preliminary process plan, is within approved economic and budgetary guidelines, a formal and very detailed process plan will be generated. This detailed process plan must specify every bit of information required to manufacture each individual part, which is often referred to as a *detail*. The process plan also must specify similar information for each assembly and subassembly being produced. Every item in the process must be decided upon. Some of these items, to name a few, are: process dimensions and tolerances, machine feeds and speeds, durable and perishable tooling, gauge inspection, and material-handling methods. The result may be a more detailed cost estimate, a routing, a series of operation sheets, or coded instructions for numerically controlled equipment. Documentation of this information will be covered in Chapter 5.

With the aid of many people, such as tool designers, tookmakers, set-up personnel, and other skilled tradespeople, the process plan will be turned into the reality of hardware out on the shop floor. This will be a time for refining and debugging the entire plan. The process plan, with all its related machinery and tooling, must work well, which means that quality parts must be produced within the original economic guidelines. Future and ongoing planning efforts will be directed toward improving the original plan and "process planning" new products. Therefore, good process planning is the economic heart and soul of any manufacturing concern. Its importance cannot be overemphasized. That is why researchers are diligently working to develop methods of improving and automating certain elements of process planning as it relates to manufacturing. These innovations will be discussed in Chapter 7.

Process planning represents the primary link between product design and shop-floor manufacturing productivity. This link and early manufacturing input in the product-design phase have been deemed so important that one major division within General Motors is moving its manufacturing engineers (process planners) into the same work section as its design engineers in order to improve communication. This better communication will aid in the designing of products that facilitate manufacturing.

Because of the relative importance of process planning to manufacturing productivity, this text has been devoted entirely to this single subject. The book can be used as a primer for students who are studying the subject for the first time or as a handbook for the seasoned professional.

1 ▪ 2 THE PROCESS PLANNER

The Process Planner Defined

The process planner is a manufacturing and process-engineering specialist who is responsible for the conception, planning, and implementation of economically justifiable production processes designed to produce a variety of industrial and consumer goods. Armed only with a product drawing and an estimate of the number of parts required, the process planner must be able to visualize several possible methods for turning the raw material into a finished product.

Applied Knowledge

To become an accomplished process planner, much applied knowledge is required. Although diverse, this knowledge will generally fall into one of the following categories.

PRINT READING

The process planner's job typically begins with an in-depth analysis of the product being considered for manufacture. The product is normally defined through one or more product drawings. These drawings can be extremely complex, and the proper interpretation of the information encountered is of paramount importance. The process planner will examine these drawings and use them as the basis for calculations. These drawings will also be used for locating points, dimensions, tolerances, material specifications, surface finishes, and all other pertinent data. Most subsequent decisions made by the process planner are directly influenced by the information extracted from these prints.

MATERIALS

Products are made from materials, of course, and in most cases these materials are specified on the product drawing. The specifications may say "make from SAE 1018 steel" or "make from cast aluminum No. 3003." Yet, the process planner must know more about the material being specified than what usually appears on the product drawing. For instance: In what forms are the raw materials available? How are they toleranced? What are the costs? What type of finish work is required? How is the material shipped and received? The answers to these questions influence, if not govern, choices made about processing methods and tooling requirements.

MECHANICAL ENGINEERING

Subjects such as the static and strength of materials, metallurgy, kinematics, fluid mechanics, and thermodynamics are aspects of mechanical engineering, but they also form the base upon which product engineering is built. ***Product engineers,*** who are also called ***design engineers,*** are usually graduates of mechanical engineering programs, so most of their thinking and decisions are grounded in this fundamental subject area.

As the process planner begins to evaluate the product design for productivity, he or she may recognize possible problems. It is important to suggest solutions or design changes that are in accord with sensible engineering reasoning and judgment. The failure of the process planner to see design-engineering problems within the product's specifications will ultimately result in a loss of revenue.

PRODUCTION PROCESSES

Process planning requires a knowledge of production processes and the ability to specify them. This includes far more than simply understanding what a casting is or what a lathe does. For example, there are more than a dozen major casting processes and seven major styles of lathes. The process planner must know all of these. He or she must comprehend the full range of existing production processes and their capabilities. This knowledge often takes years to master. The appendixes of this book have been assembled to expedite the initial learning process and to make future reference easier.

MACHINERY (CAPITAL EQUIPMENT)

In process planning, proper machine selection is based upon the machine's capability to perform a given task. Dimensional accuracy and repeatability, cycle time, flexibility, cost, and cost-to-tool are all items that affect the economics of machine selection.

Here again, the process planner must have or have access to an extensive knowledge base. This knowledge base will include information about most existing machine tools. Such data will help to ensure the specification of the best overall machine for each planned operation.

TOOLING AND TOOL DESIGN

Technically, there is a distinction between a machine tool and its tooling. The typical machine tool (e.g., a drill press, lathe, milling machine, grinder, etc.) is designed to be capable of accommodating a certain type of work, over a wide range of sizes. Tooling, in turn, allows a machine tool to perform specialized operations on specific parts. Tooling is further classified as either durable or perishable. Durable tooling is the portion of the setup that will last for years under normal operating conditions in industry. Perishable tooling is the portion of the setup that will wear out and require replacement over and over again.

Examination of three typical industrial processes will clarify the differences between machine tools, durable tooling, and perishable tooling:

1. A drilling process requires a drill press (machine tool), a drill jig (durable tooling), and a drill bit (perishable tooling).
2. A turning process may require a lathe (machine tool), a chuck (durable tooling), and a cutting tool (perishable tooling).
3. A grinding process requires a grinder (machine tool), a part-holding fixture (durable tooling), and a grinding wheel (perishable tooling). (Curtis, *Tool Design,* 1986)

The process planner must understand these distinctions in tooling and become familiar with what is commercially available.

Required tools that cannot be purchased must be designed and built. It is the process planner who specifies all necessary tooling, whether purchased in standard format or specially designed by tool designers.

TOOLROOM PRACTICES

After tool designers draw up a plan for the tooling, toolmakers, using standard toolroom practices, turn the plan into hardware. Every decision made by the tool designer about the design configuration and its tolerances must include a consideration of precisely how each detail is to be made. (Curtis, *Tool Design,* 1986)

Because process planners request and approve tool designs, they too must have an intimate understanding of standard machining and toolroom practices.

AUTOMATION TECHNIQUES

As previously stated, the process planner is required to specify the machines and tooling necessary to manufacture a given part. However, beyond the basic needs of the operation, the degree to which a process is automated becomes a management decision that is first recommended and justified by the process planner.

The availability of inexpensive and unskilled labor has always offered management a low-risk option requiring little or no capital investment. Therefore, the process planner must become skilled at recognizing automation techniques that minimize risk as they enhance manufacturing productivity.

INSPECTION TECHNIQUES

The variability of all manufacturing processes and the demand for parts that are dimensionally correct leads to the necessity of production gauging-and-inspection techniques. In addition to specifying a method of manufacture, the process planner must indicate how the dimensions and specifications of each part are to be inspected. It is also important to indicate what means will be used in making such inspections.

STATISTICAL PROCESS CONTROL

Statistical process control is based upon the premise that there is variability in every process. Through the use of statistical methods, accurate predictions can be made about the extent of this variability. By inspecting a small number of parts, called a *sample,* the process planner can draw statistical conclusions about a larger number of parts or about the capability of a machine and its tooling. This concept of statistical process control can be used to test and prove new machinery. It can also be used in monitoring shop-floor quality at a minimum cost. Therefore, a knowledge of statistical process control goes hand in hand with that of inspection techniques.

ENGINEERING ECONOMICS

There are many ways to produce a given part, but the process plan selected should yield the lowest per-unit cost. Raw material, direct and indirect labor costs, overhead, production volumes, product stability—all of these factors impact the final decision. However, these factors are only part of the story.

A good process planner must bear in mind the old saying "It takes money to make money." Loosely translated, this means that money must be spent today in hopes of achieving a real dollar return on the investment in the future. Accurately predicting the future return of money spent on engineering, design, machinery, and tooling requires *engineering economics*.

At the foundation of engineering economics is a concept called the *time value of money*. For example, if $500,000 is spent today to start manufacturing a product, how much money should be returned two years from now to make it a good investment? Before answering too quickly, consider inflation, the loss of interest, risk, equipment depreciation, tax laws, and the method of repayment. These factors complicate the matter. Therefore, a good process plan has to make the part as well as a profit, and that requires an understanding of both engineering and economics.

DRAFTING AND COMPUTER-AIDED DESIGN

Drafting, the ability to develop and construct a readable drawing in accordance with industrially accepted symbols and practices, is a fundamental skill required of the process planner. This skill includes the drawing of process sheets, a responsibility that is often given to the planner (see Chapter 5.5).

In time, computer-aided design will render traditional lead-on-paper drafting techniques obsolete, but this change in drafting media will be a gradual one. Therefore, a good process planner will need to know both manual and computer-aided drafting techniques.

GROUP TECHNOLOGY AND CAPP

In most manufacturing concerns, it is possible to group a variety of parts according to their similarities in design and manufacture. Grouping diverse parts into families makes it easier to accommodate the processes and tooling designed for one part to other parts within that family.

The computer is useful in keeping track of existing production parts, their attributes, processes, machinery, and tooling. When it is necessary to make new parts, the planner can refer to information stored in the computer's memory in order to find out about existing processes and tooling. The use of the computer in this fashion is called *Computer-Aided Process Planning* (CAPP). Group technology and CAPP, which are more fully explained in Chapter 7, have major implications for the process planner.

Specialization

After acquiring the necessary basic planning skills, the process planner will normally specialize in one or more of the following areas:

1. Machining
2. Pressworking (Forming)
3. Plastics
4. Finishing and Coating
5. Welding and Adhesives (Fabrication)
6. Assembly
7. Other Highly Specialized Areas

This specialization takes place for two different but related reasons. First, factories that employ process planners often specialize in products that dictate the planning of a specific type of process. Second, the breadth of the field necessitates specialization as a prerequisite for mastery.

Education

Considering the applied knowledge required of the process planner, some combination of formal engineering training and industrial experience offers the best overall education.

Today, however, for a variety of reasons, most industries require a bachelor of science degree in engineering or engineering technology for entry-level positions in process planning. Individuals hired for these entry-level jobs usually are called *process engineers* or *manufacturing engineers*. As with most technical professions, these jobs require more formal education than in the past. Because the process engineer must remain technically up-to-date, education must be considered a lifelong process. Additional college course work, industrial seminars, membership in professional organizations, review of trade magazines, and independent study are all excellent ways of keeping up with the ongoing information explosion.

Manufacturing Engineering

In any discussion of process planning, the larger sphere called manufacturing engineering must be examined. In general, any of the functions or departments involved in putting a product into production could be considered aspects of manufacturing engineering. However, the following subsections will give a more specific idea of what this function entails:

1. Manufacturing Liaison
2. Process Engineering
3. Methods Engineering
4. Tool-and-Machine Design
5. Tool-and-Die Making
6. Tool Control
7. Plant Layout

MANUFACTURING LIAISON

This group is charged with communicating information between product design and process engineering. Manufacturing liaison advises product design about the manufacturability of an idea and keeps process engineering apprised of plans for future products, innovations, and changes.

PROCESS ENGINEERING

Industry has chosen process engineering as the title for the function of process planning as covered in this text.

METHODS ENGINEERING

This group identifies and plans the exact motions required to complete a given operation efficiently. This involves making layouts for the workplace, establishing time standards, writing job-instruction sheets, and identifying all other operational parameters that ensure the minimizing of manufacturing costs.

TOOL-AND-MACHINE DESIGN

This group develops and documents the tooling and special machinery required for all phases of manufacturing.

TOOL-AND-DIE MAKING

This skilled-trades group actually builds the tool, dies, jigs, fixtures, and special equipment designed by the tool-and-machine designers.

TOOL CONTROL

This group is responsible for getting, storing, and dispensing the tools required to sustain manufacturing within the plant.

PLANT LAYOUT

This group plans the arrangement of equipment and other support services within the plant to facilitate the efficient manufacturing of products.

1 ▪ 3 ECONOMIC IMPLICATIONS

At this point, the economic importance of process planning (process engineering) to industry should be evident. The quality and accuracy of the preliminary process plan (cost estimate) is inextricably connected to the economic well-being and the very survival of the typical manufacturing concern.

If the original estimate of manufacturing costs is inaccurately high, one of two things will occur: (1) the independent manufacturer may decide against entering what could have been a very lucrative market, or (2) the manufacturing job shop will be the high bidder and, thus, uncompetitive in its quoting against other job shops.

If the original estimate of manufacturing costs is inaccurately low, companies will enter markets and win bids only to find that they are losing money with every part produced.

Both errors will spell economic disaster.

Although other circumstances can contribute to the demise of a company, no other single function can compare to process planning in its effect on overall manufacturing profitability.

REVIEW QUESTIONS

1. What is process planning and what is a process plan?
2. How does the process planning effort begin?
3. Preliminary process plans are sometimes referred to as cost estimates. How does management use these so-called cost estimates?
4. What is the link between product engineering and shop-floor productivity?
5. Who is the process planner?
6. What kinds of applied knowledge is required of a process planner?
7. Why do most process planners specialize?
8. What kind of formal education is required of the process planner?
9. Two separate but related problems are caused by inaccurate process plans; what are they?
10. What single area has the greatest effect on overall manufacturing profitability?

CHAPTER 2

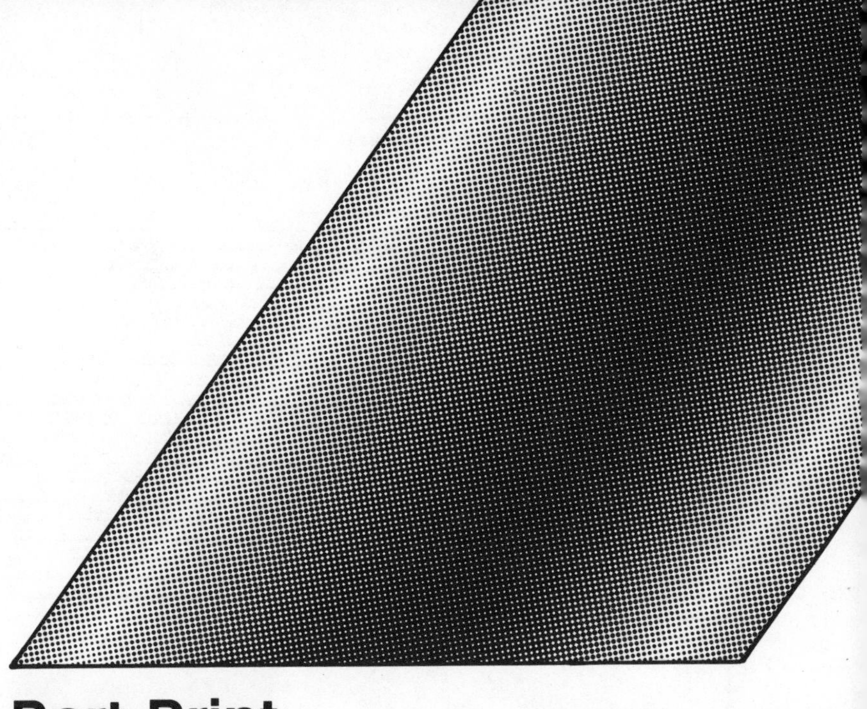

Part-Print Analysis

2 ▪ 1 INTRODUCTION

In the typical manufacturing company, process planning begins with an in-depth analysis of the product drawings, which are often called *part prints*. These part prints may include a parts list and drawings for assembly, subassembly, and detail as required (see Figs. 2-1, 2-2, 2-3, and 2-4, respectively).

This analysis is normally triggered by a formal request for cost estimate (see Fig. 2-5). The *cost estimate transmittal* (CET) provides the process

PARTS LIST
16.5 * 7 BRAKE ASSEMBLY

DRAWING SIZE	DRAWING NO.	NUMBER REQUIRED	DESCRIPTION
D	20001	2	Shoe & Lining S/A
D	70001	2	Shoes
C	90031	4	Lining Blocks
B	90002	40	Rivets
D	40001	1	Air Chamber Mtg. Bracket S/A
D	70002	1	Air Chamber Mtg. Bracket
A	90021	2	Seals
A	90022	2	Bushings
D	60001	1	S Cam Shaft
C	80001	1	Shoe Retaining Spring
C	80002	2	Anchor Spring
B	60011	1	Anchor Pin
E	30001	1	Spider
C	50001	2	Rollers
A	90001	4	Mtg. Bolt
A	90011	4	Nut

ECN	LTR	CHG	DATE	BY	CK.	SHT. 2 OF 2 / NO. 10001-PL

Figure 2-1 Parts List

planner with estimated sales data, customer name, part or drawing numbers, and other pertinent information relating to the required accuracy and urgency of the estimate. This document is covered in greater depth in Chapter 5.2.

Armed with a CET and its corresponding product drawings, the process planner must embark upon a complex decision-making path. This path is outlined and explained in a step-by-step fashion in this chapter and in Chapters 3 and 4. If properly followed, the steps along this path will result in a workable and highly cost-effective process plan.

2 ▪ 2 MAKE-OR-BUY DECISIONS

A company's decision to make some components and buy others is fundamentally economic in nature. The process planner makes this decision, which is the result of research and an analysis of each part print. This re-

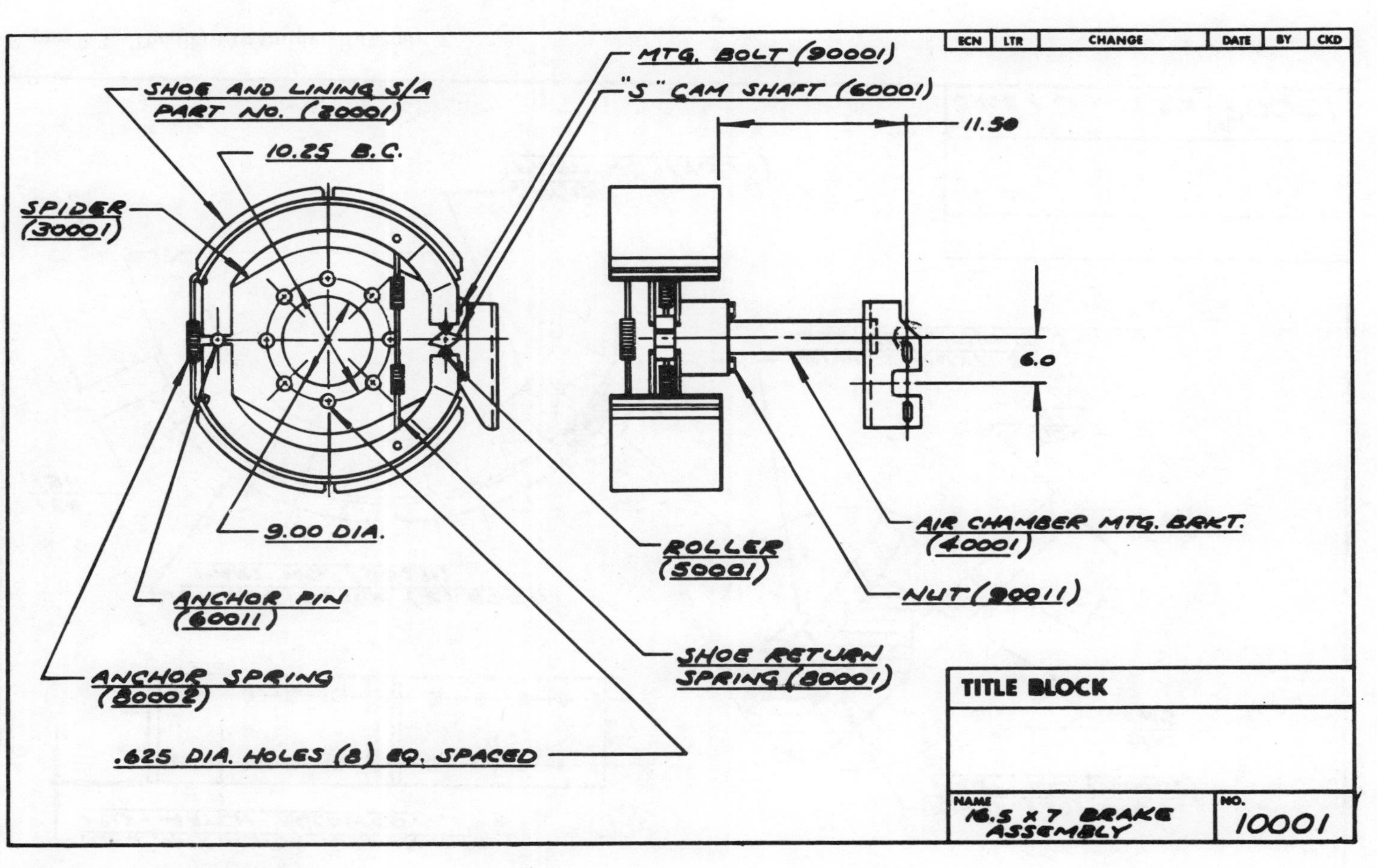

Figure 2-2 Assembly Drawing

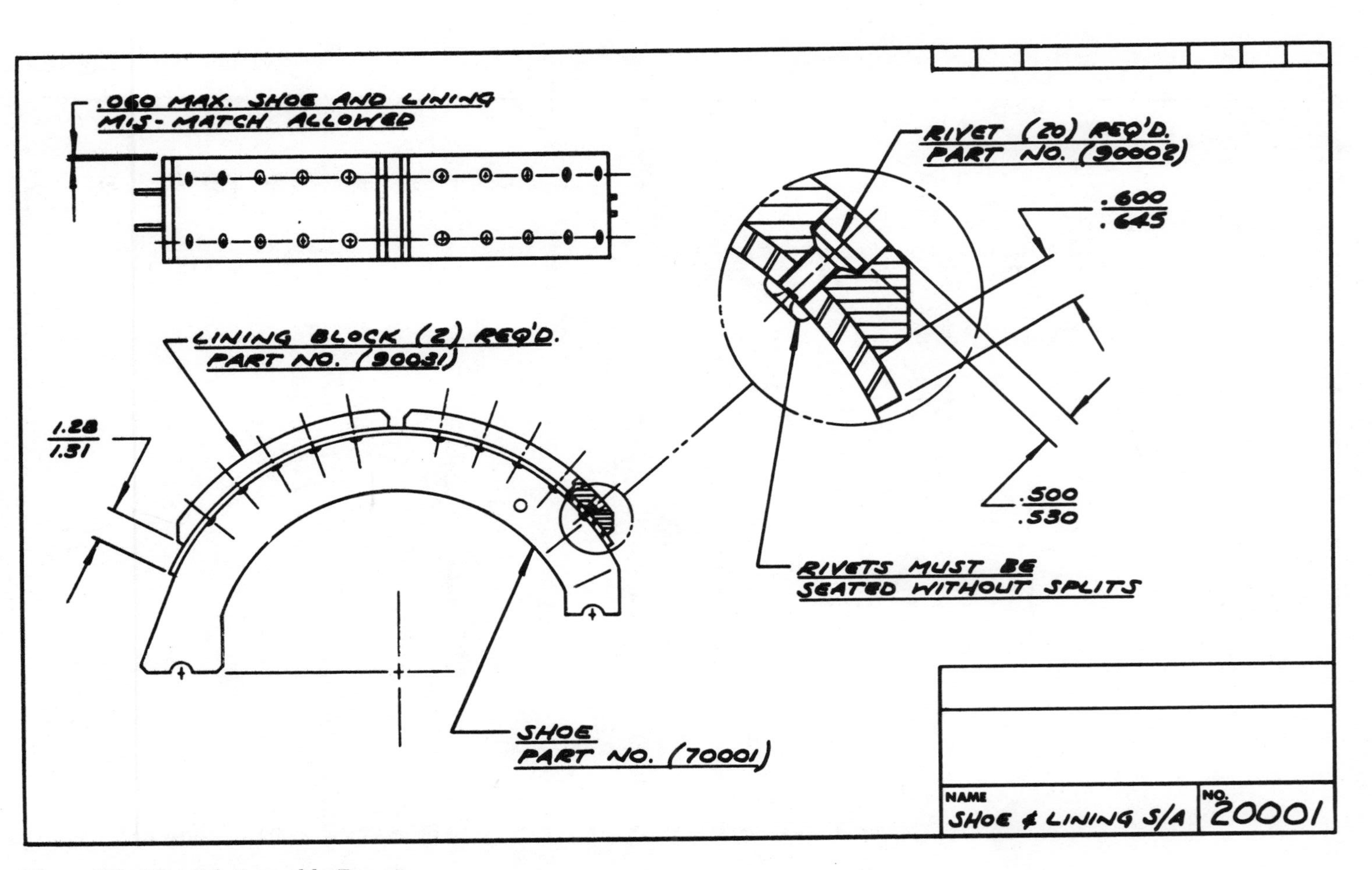

Figure 2-3 The Sub-Assembly Drawing

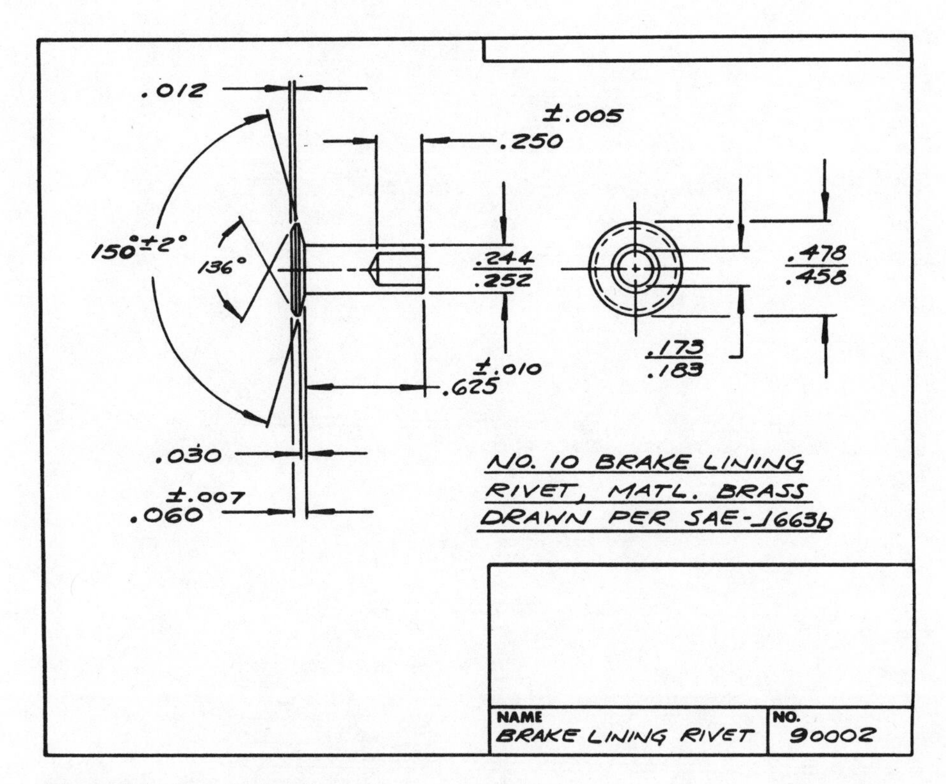

Figure 2-4 Detail Drawing

search can eliminate wasted engineering effort on components that will ultimately be purchased.

Take, for example, an automobile built by General Motors. It could contain a differential manufactured by Eaton Corporation, piston rings from Sealed Power, wheels made by Motor Wheel and a carburetor produced by Holley Corporation. Although the average consumer might think of the car as a product manufactured entirely by a single corporation, many of the components could be produced for General Motors by a variety of automotive suppliers. Such a supplier is called an original equipment manufacturer (OEM). In this case, it would be safe to assume that the process planners at General Motors made a bottom-line, dollars-and-cents decision to buy, rather than make, some of the components required in the production of a completed automobile.

The following subsections outline the specific items that must be considered by the process planner to complete intelligently what industry calls the make-or-buy decision.

COST
ESTIMATE
TRANSMITTAL

CET NO.________________
DATE OF ISSUE__________
DATE REQUIRED__________

DATA REQUIRED

☐ FIRM PROCESS ☐ PRELIMINARY PROCESS
☐ FIRM MATERIAL ☐ PRELIMINARY MATERIAL
☐ ADD-DELETE ☐ OTHER
☐ NEW ☐ REPEAT BUSINESS

DATA KNOWN

☐ CUST. EST. ☐ SALES EST.
LOT SIZE______ PACK SIZE_____
PER MO.______ ANNUAL________
PROD. TARGET DATE___________
REPLACES PART NO.___________
DRAWINGS AVAILABLE YES NO

SALES COMMENTS (SALES ENGINEERING)

MANUFACTURING COMMENTS (PROCESS ENGINEERING)

LEAD TIME FROM FORMAL PROGRAM RELEASE TO:
1) PRODUCTION READY
2) DELIVERY OR COMPLETION IN HOUSE

CAPITAL EQUIPMENT	______WKS.	$________
VENDOR TOOLING	______	________
IN PLANT TOOLING	______	________
GAGES & Q.C.	______	________
SETUP	______	________
TOTAL	______WKS.	$________

CC: GENERAL MGR.
MFG. MGR.
MKTG. MGR.
ADV. ENG MGR.
CH. PROD. ENG.
CH. MFG. ENG.
SALES MGR.
MATERIALS MGR.
PRODUCTION MGR.
PURCHASING MGR.
CONTROLLER
Q.C. ENG.

POSITION	SIGNATURE	DATE
PREPARED BY		
PROC. ENG.		
Q.C. ENG.		
CH. MFG. ENG.		
MFG. MGR.		

Figure 2-5 The Cost-Estimate Transmittal

Buy Considerations

First, is the part (or something very similar) commercially available? If yes:

1. Is the price reasonable?
2. Is the quality acceptable?
3. Can your production quantities be met?
4. Can it be obtained from at least two sources—that is, can it be dual-sourced?

In some cases, these questions can be answered only after a detailed analysis of in-house manufacturing costs. This could mean that a simplified process plan would need to be completed prior to formulating the make-or-buy decision. However, if it is possible to answer yes to the aforementioned questions, the part being analyzed is a good candidate to be purchased on the outside.

It may also make sense to purchase the components as opposed to making them if volume levels are extremely low or if the potential market for this product is questionable. By purchasing components, the process planner minimizes risk, keeps capital unencumbered, and pushes both the cost of inventory and quality on to the supplier (also referred to as the *vendor*). In fact, Chrysler Corporation is working its way toward *out-sourcing* or purchasing all the components necessary to produce an entire automobile.

Another reason often cited for out-sourcing of component parts is short lead times. For example, a company may be required to bring a given product to the marketplace in just 12 weeks to maintain its competitive edge. Suppose it would take 20 weeks to purchase and tool a machine capable of producing the product in question. In such a case, it would be necessary to purchase the component parts of the product, even at a premium price, in order to meet the short lead time of 12 weeks. Out-sourcing would at least be a temporary solution.

In any event, the purchasing of component parts from outside suppliers must be considered by the process planner.

Make Considerations

In considering the desirability of making (manufacturing) a component in-house, the process planner must ask the same series of questions as during the buy decision. In this case, negative responses would lead the process planner towards in-house manufacture of the needed component.

When there is not a clear answer to the make-or-buy question, there are other factors that may lean towards a make decision. For example, if the part is made in-house, any of the following could result:

1. The company gets all the profit.
2. The company has control over price, quality, and delivery.
3. The fixed overhead cost of the plant receives additional support (see Fig. 2-6).
4. The tooling costs will be the same because the company must pay for any special tooling required whether the components are made in-house or outside.

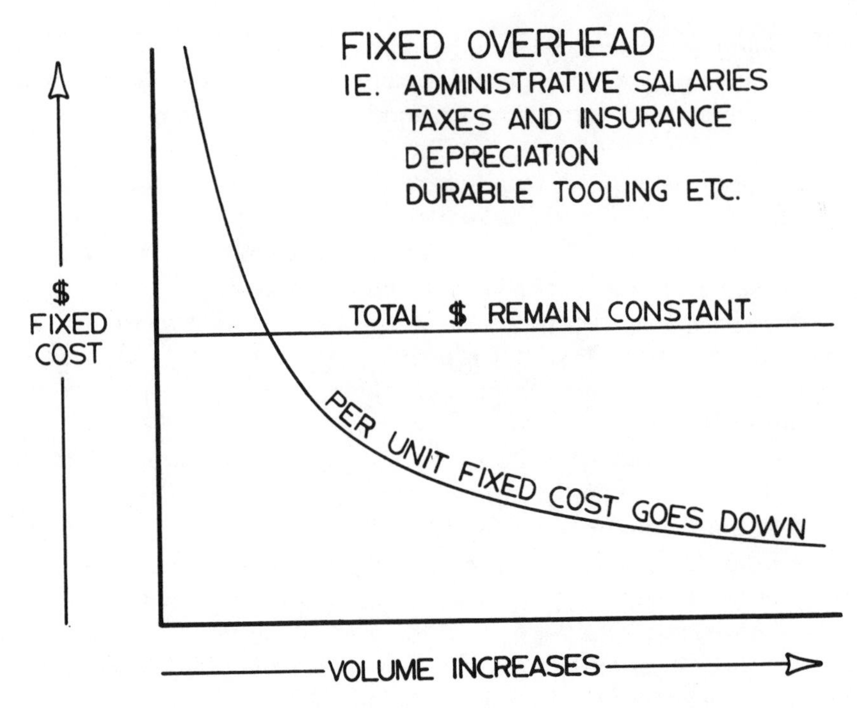

Figure 2-6 Fixed Overhead

Make-or-Buy Decision Completed

Upon the completion of the make-or-buy decision, the process planner should have two stacks of prints: one labeled *Make* and the other labeled *Buy*.

The *Buy* stack of prints must be sent to the purchasing department along with information estimating costs, part quality, quantity, and delivery requirements; this stack should also include a list of potential suppliers. The purchasing department will take over at this point by soliciting quotes from appropriate vendors and assigning formal purchase orders.

The *Make* stack of prints is kept by the process planner, who may now continue to the next step in the part-print analysis procedure (see Fig. 2-7).

2 ▪ 3 RAW MATERIAL SELECTION

At this point in the analysis of the part print, two things should be known:

1. The final form of the parts or assemblies; that is, dimensions, tolerances, and so forth.

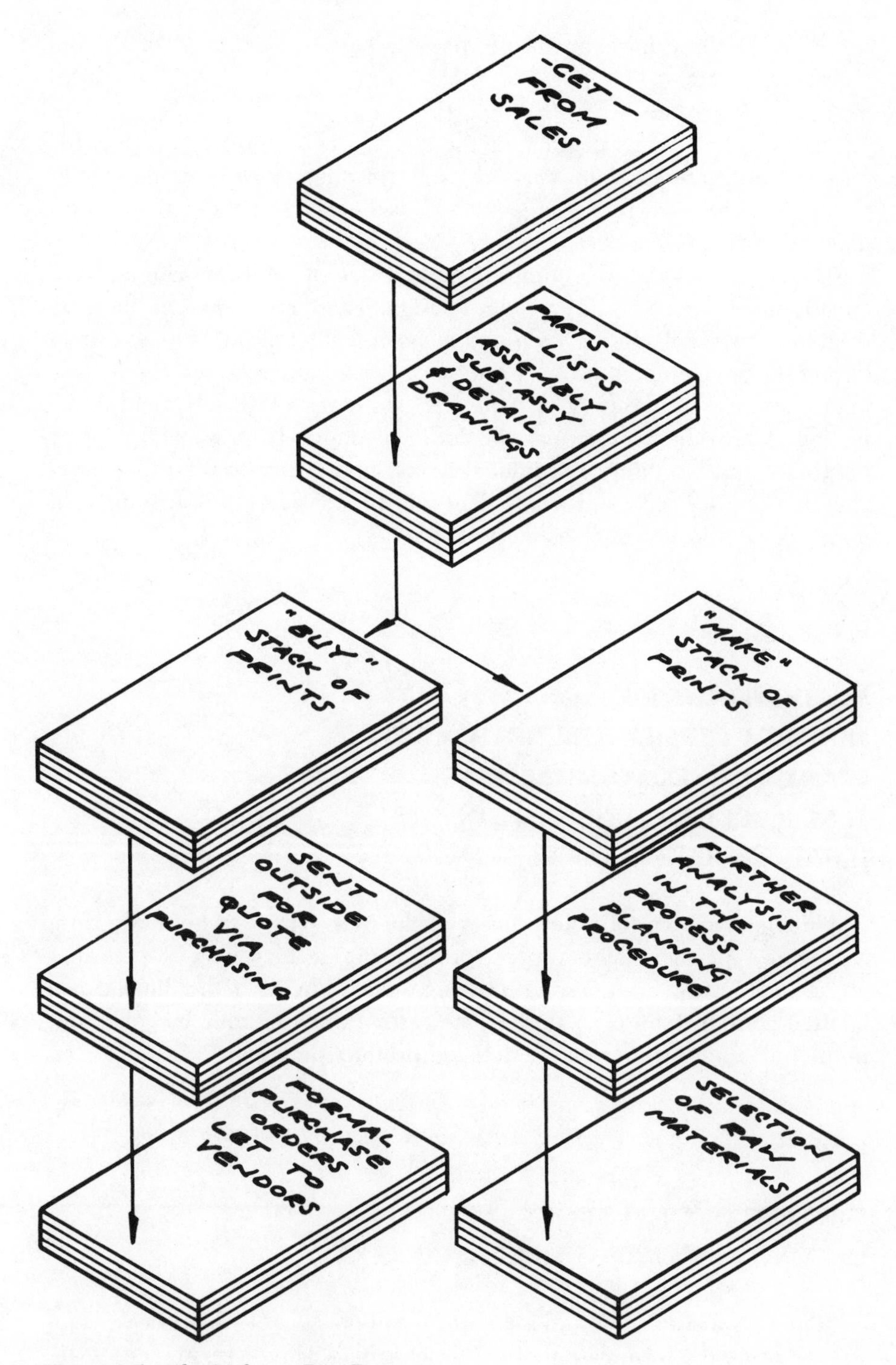

Figure 2-7 The Make-or-Buy Decision

2. Whether those parts are to be made in-house or out-sourced as purchased items.

Inasmuch as process planning is the creating of a step-by-step plan for turning raw materials into finished products, the selection of these raw materials for all parts to be made in-house is the next step along the process planning path (see Fig. 2-7).

In some cases, both the composition and size of the raw material have already been established by product design. These are shown on the part print as a *material specification* and a *make-from* call out. For example, during the development of most products, a company may test the performance and use of a variety of materials. These tests typically yield information about whether the materials are acceptable from a standpoint of durability and reliability. Additional items such as material cost, appearance, processing cost, and availability will be considered in arriving at a material specification such as the following.

MATERIAL

SAE D5506 DUCTILE IRON
MINIMUM TENSILE STRENGTH: 80,000 PSI
MINIMUM YIELD POINT: 55,000 PSI
MINIMUM ELONGATION IN 2 IN.: 6%
HARDNESS RANGE: 179–255 BHN

This specification tells the process planner the part is to be made from a specific type of cast iron. Here, product design may go a step further and actually draw up a *casting drawing,* which identifies the dimensional configuration of the raw material. The casting drawing may be called out in the following manner on the detailed product drawing.

MAKE FROM

SAE D5506 DUCTILE IRON
CASTING DRAWING NUMBER: XXXXXXXX

Material specifications and make-from call-outs enable the process engineer to begin the formulation of the process plan. However, in many manufacturing situations, the specification of the raw materials is left largely up to the processing group.

Again, performance and economy become the primary factors affecting the material-selection decision. A series of material-selection tables (Appendix II) have been prepared and placed in this book as an aid to the process planner facing such a question.

Whether the material specifications come from product design or process engineering, the raw material must be specified prior to further planning of the manufacturing process.

2 ▪ 4 PRODUCT-DRAWING ANALYSIS

The complete analysis of a complex product drawing is a necessary yet difficult proposition. The key words here are *complete analysis*. By simply overlooking or misinterpreting a single tolerance or note, the process engineer could develop a process plan that is either inadequate or incorrect. Faulty process plans turn into economic disaster; they often result in the addition of costly operations, reworking, or a quality-related loss of business.

To eliminate the problems caused by such oversights, the process planner must develop a systematic method of product-drawing analysis. What follows is an example of one such systematic plan.

First, run prints of all product drawings to be reviewed. Then, one by one, make a general review of each print for critical dimensions, clamping- and locating-points, and specific process requirements. Highlight tight tolerances, datum surfaces, and notes calling for specific processes such as paint or heat. Next, use the letters and numbers shown outside the borderline of the print to establish a gridwork (see Fig. 2-8). Now, each square within the grid can be studied as was the entire print. These smaller areas help to focus and organize the print review procedure; they minimize the possibility of costly oversights. Keep primary notes in reference to and identified by a specific grid location on the print, such as A-1, A-2, A-3, and the like. This will simplify further work on or reviews of the process plan. Keep all work completed at this stage—prints, notes, CET, and so forth—in appropriately labeled file folders.

Critical Dimensions

As the process planner makes a dimensional review of the product drawing, three questions should be asked:

1. Are there dimensions and/or views missing in the graphic description of the part's size and shape?
2. Are the associated dimensional tolerances unrealistically tight?
3. Are there obvious errors in the dimensioning and tolerancing of the part?

If the answer to any of the above questions is yes, the process planner must review the items in question with the product-engineering department. This review could lead to the processing of a formal *engineering change notice* (ECN), which is fully covered at the end of this section.

Once these preliminary questions have been resolved, it is time to zero in on the *critical dimensions*. A critical dimension has importance or significance. These critical dimensions can be identified by applying the following questions:

1. Does this dimension and tolerance imply the use of a certain process (see Appendix III)?

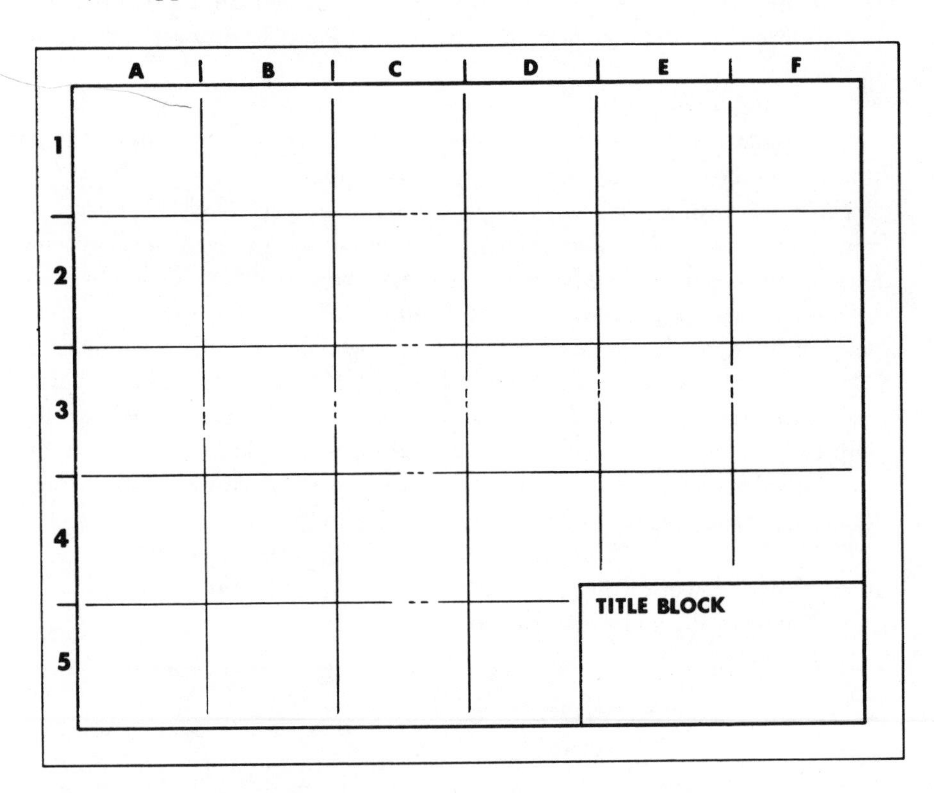

Figure 2-8 Print with the Gridwork

2. Is this dimension or feature to be used as a datum (namely, a locating point) in further processing of the part?
3. Will this dimension require special material-handling techniques once completed?
4. Will there be special inspection methods required for this dimension and tolerance?

Once identified, the critical dimensions become the focal point of the process-planning effort.

The Engineering Change Notice

An Engineering Change Notice (ECN) is a formal document that authorizes a change to be made on an existing product drawing (see Fig. 2-9). It states, in both word and drawing form, exactly what is to be changed and

ECR ______ ENGINEERING CHANGE ECN ______

x	Reason For Change	DETAILS	Date
	Decrease Cost		Origin
	Improve Quality		Part No
	Correction		Name
	Customer Request		Customer
	Mandatory		Dwg Chg Ltr
	Facilitate Mfg		Supersedes
	Other		Service Aff
			Stock Disposition
			Approved By

Process Remarks	Cost	Pres	Prop	Diff
	Material			
	Labor			
	Burden			
	Total			
	Tools			
	Equip			

Production Remarks

With Obsolescence

Without

Sales Remarks	ECN	COPY	INT
		Prod Mgr	
		Mfg Eng	
		Sales Mgr	
		Sales Eng	
		Superintendent	
		Eng Mgr	
Effective	Issued By	Product Eng	
Date Issued	Approved By	Originator	

Figure 2-9 The Engineering Change Notice

for what reasons. Although the ECN is not actually a product drawing, it affects product drawings to such an extent that a review here is necessary.

Once any product design is released to sales, purchasing, and manufacturing engineering, it is considered to be in final form. This final form simply means that one workable design solution has been agreed upon. For many reasons, a design change may be required after the design is released to production. Some of the more common reasons for these design changes are listed below:

1. To correct an error
2. To facilitate manufacture
3. To improve product performance
4. To minimize product cost
5. Customer request
6. Standardization
7. Safety or product liability

Many times a single design change will have far-reaching implications in terms of the time and cost required for implementation. When this is the case, a document referred to as an Engineering Change Request (ECR) precedes the generation of an ECN. The ECR must be approved by manufacturing engineering and other affected departments before an ECN becomes formal.

Much of the manufacturing engineering and tool design work completed for existing products is originally generated by the ECN. The importance of the ECN cannot be overemphasized and the manufacturing engineering and tool design departments must give high priority to the completion of all work dictated by the ECN. (Curtis, *Tool Design,* 1986)

Clamping- and Locating-Points

The process-planning effort now requires identification of clamping-points and, more important, the locating-points on the workpiece. Proper identification of these points is essential in maximizing, or effectively using, the dimensional tolerances set forth on the product drawing.

DATUM

In companies using geometric dimensioning and tolerancing, prospective locating-points will be specified on the product drawing as *datums* (see

Fig. 2-10). In practice, the datum provides a reference for the calculation, manufacture, and measurement of part features under geometric or dimensional control. A datum can be established by a point, line, plane, cylinder, or axis. Any part feature has the potential to become a datum. When properly selected, the datum should provide a basis for the functional relationship between part features. Datums are identified with two symbol styles (see Fig. 2-11). Once identified, up to three separate datums can be referenced in a single, geometric, feature-control symbol. Figure 2-12 shows the applied meaning of these primary, secondary, and tertiary datums.

Datums are normally established and placed on product drawings prior to any process-planning or tool-design activity. This fact gives advance warning to the process engineer and tool designer; it also provides knowledge of how the part must be held, processed, and gauged.

SIMPLIFIED TOLERANCE CHARTING

In the absence of prior datum identification, the process engineer must use another method to help determine part features that are suitable for use as locating-points (i.e., datums). One such popular technique used in this regard is referred to as simplified tolerance charting (see Fig. 2-13). This is how it works:

1. Select the product drawing to be analyzed for locating-points (see Fig. 2-14).

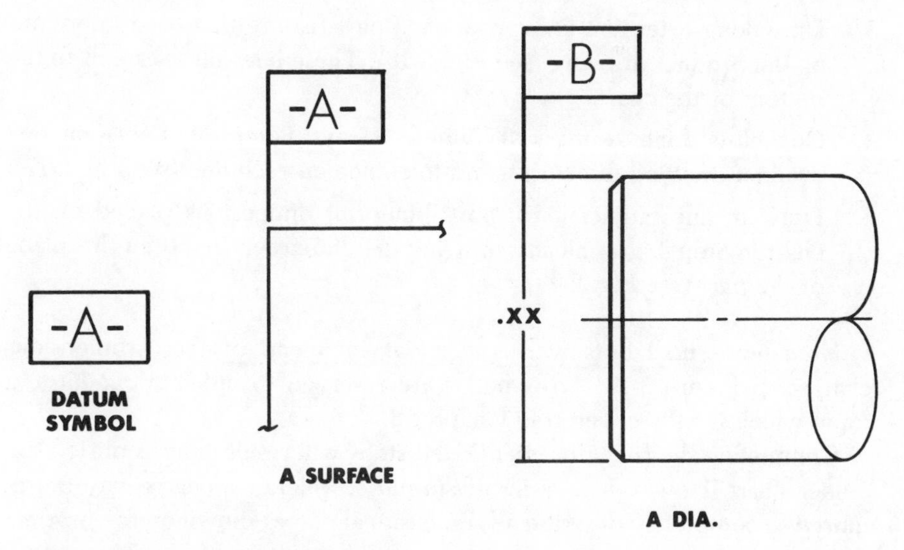

Figure 2-10 Datum Identification

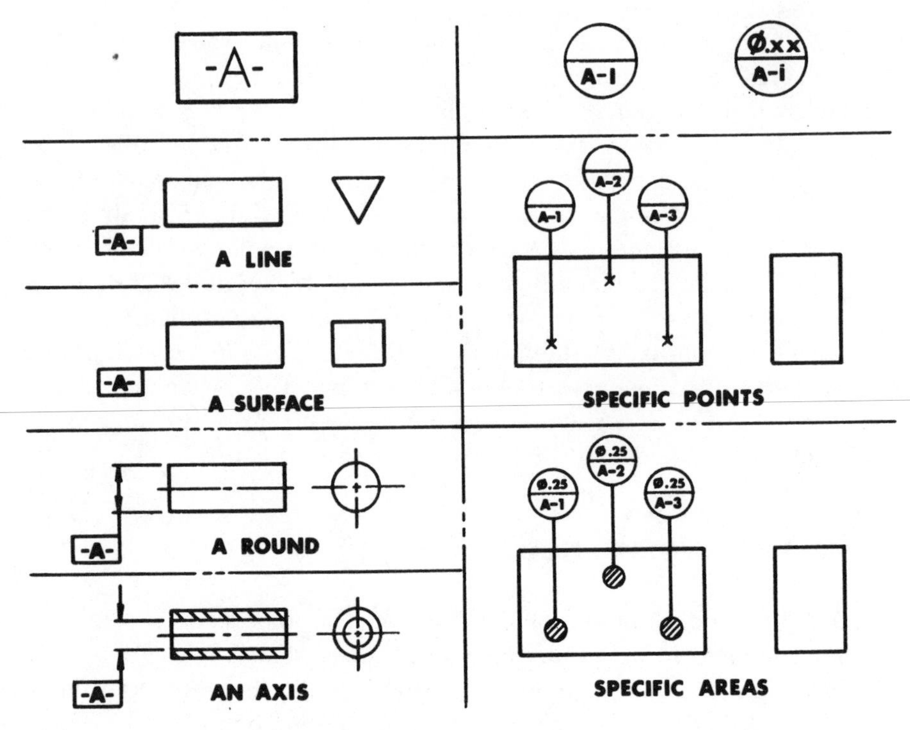

Figure 2-11 Datum Symbol Styles

2. Draw a partial view of the product at the top of a blank tolerance chart (see Fig. 2-15).
3. Draw long extension lines from each part feature that has dimensions on the product drawing (see Fig. 2-16). These lines must extend to the bottom of the chart.
4. Go below Line Number 36 where it says *Blueprint.* Pencil in one product-drawing dimension and tolerance on each line (see Fig. 2-17).
5. Draw in lines adjacent to those blueprint dimensions placed on the chart in Step 4, graphically showing the dimension found on this plane of the part (see Fig. 2-18).

Note how small dots were placed at each end of these dimensions charted in Figure 2-18. Arrowheads are reserved to indicate *cut lines,* a topic which is fully covered in Chapter 6.

Completing the five aforementioned steps will result in a simplified tolerance chart. However, a *series* of simplified tolerance charts may be required to completely describe all dimensional planes shown on the product drawing.

The process planner may now simply count the number of dots found on each extension line (see Fig. 2-19). The larger the number of dots, the greater the likelihood of using that part feature as a critical locating-point. These surfaces then, in effect, become datums upon which the process plan can be built.

Specific Processes

The last bit of information that can be directly extracted through part-print analysis (product drawing) relates to specific processes. For example, a product drawing may specify the part to be phosphate-coated and then painted. These specifications actually prescribe two processes that the planner must incorporate into the process plan. The act of incorporating these processes is called *routing*.

Another product drawing may specify a certain type of heat treatment, which also translates into specific processing requirements. Therefore, simply by reading the specifications noted on the product drawing, the process

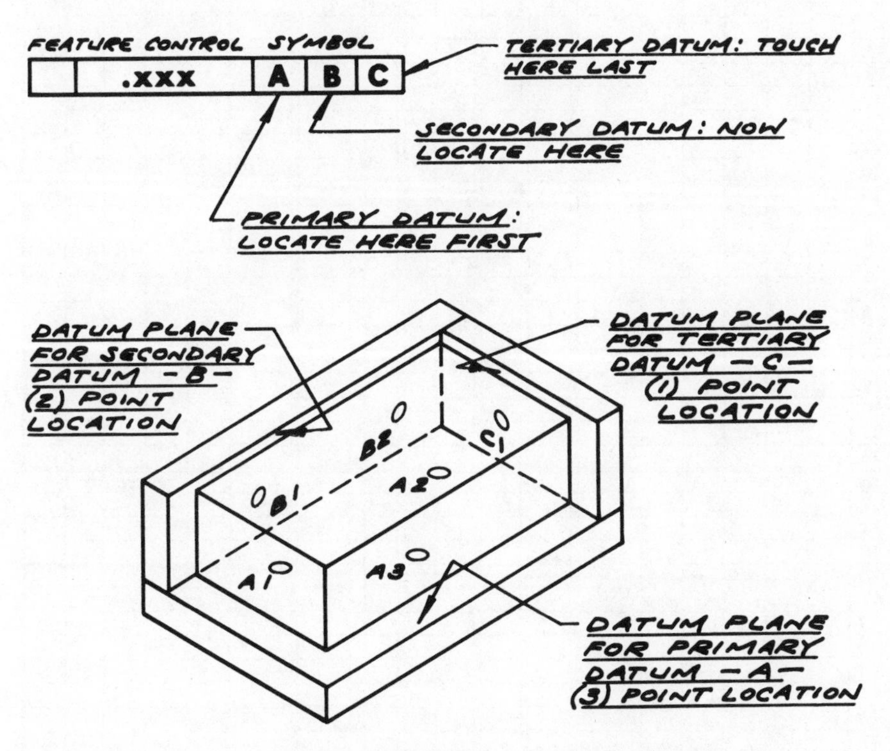

Figure 2-12 Primary, Secondary, and Tertiary Datum Explanation

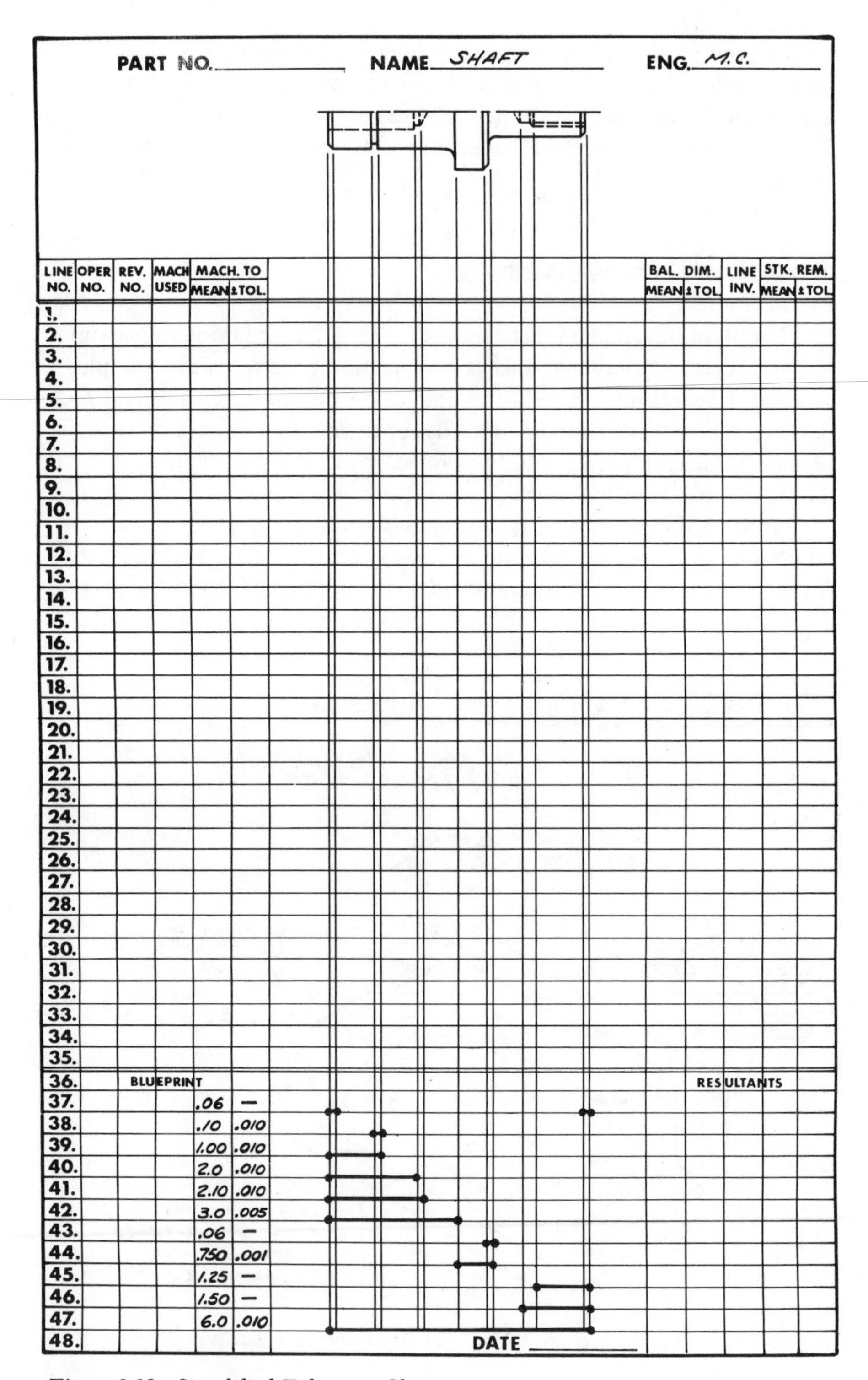

Figure 2-13 Simplified Tolerance Chart

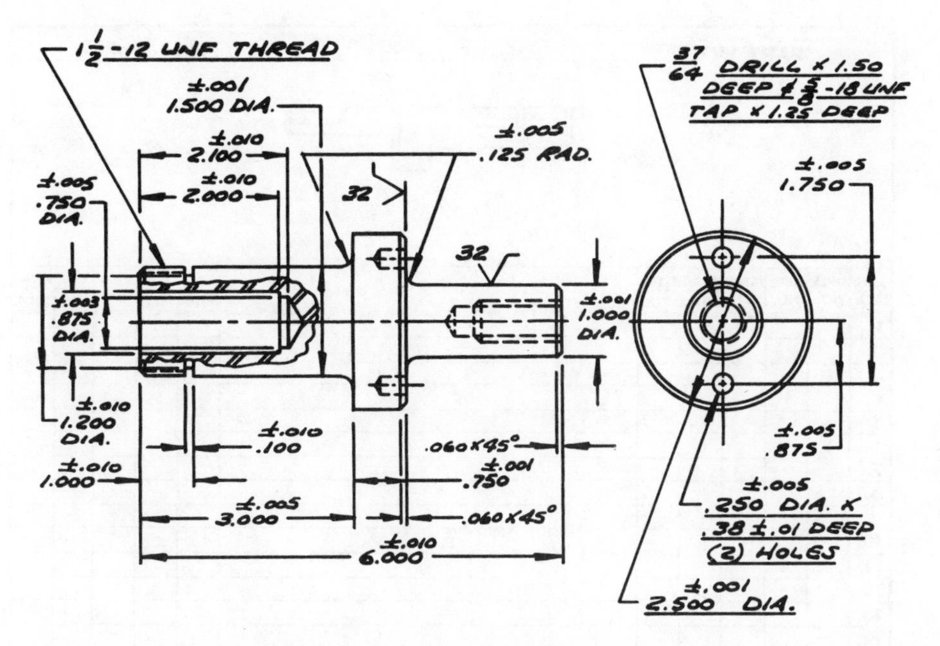

Figure 2-14 Sample Product Drawing

planner can pinpoint a number of required processes. A list of these specific processes should then be made for future integration into the process plan.

PROCESS SELECTION

Production processes must now be selected to accommodate the unique requirements of the part under analysis. These processes should be in keeping with projected production volumes, including those datums, critical dimensions, and specific processes previously identified.

To aid the planner with process selection, a series of process tables have been created and placed in this book (see Appendix III). These tables are listed alphabetically by generic or general-processing need; for example, Drilling. Listed under each generic process title is a number of specific processes, each capable of meeting the general need in some form (see Fig. 2-20). Typical applications, capabilities, and recommended production volumes will also be given for each specific process.

For those unfamiliar with some of the processes referenced within the process tables, a glossary of terms has also been provided (see Appendix I).

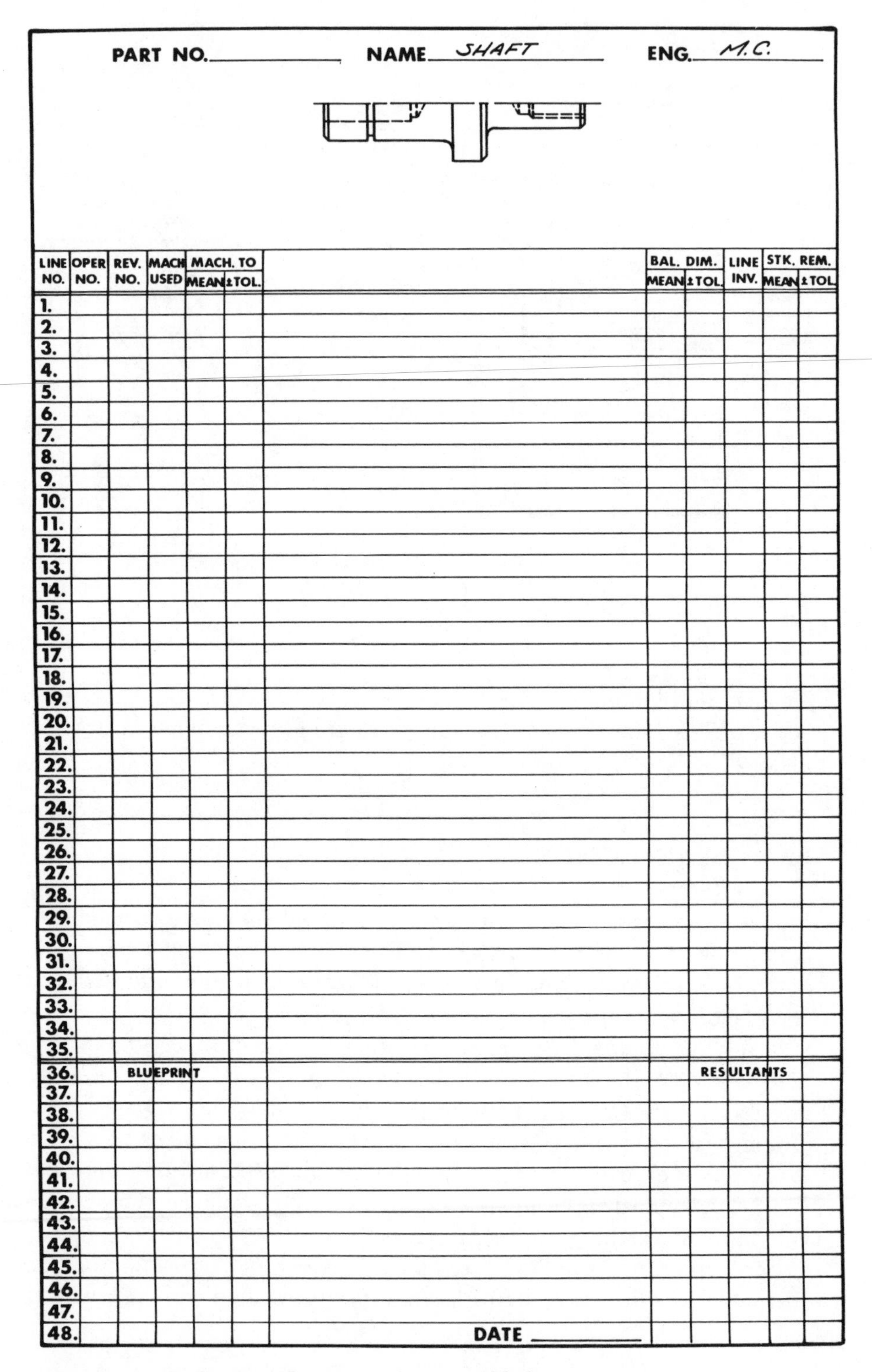

PART NO. ______ NAME SHAFT ENG. M.C.

LINE NO.	OPER NO.	REV. NO.	MACH USED	MACH. TO MEAN	±TOL.		BAL. DIM. MEAN	±TOL.	LINE INV.	STK. REM. MEAN	±TOL.
1.											
2.											
3.											
4.											
5.											
6.											
7.											
8.											
9.											
10.											
11.											
12.											
13.											
14.											
15.											
16.											
17.											
18.											
19.											
20.											
21.											
22.											
23.											
24.											
25.											
26.											
27.											
28.											
29.											
30.											
31.											
32.											
33.											
34.											
35.											
36.		BLUEPRINT							RESULTANTS		
37.											
38.											
39.											
40.											
41.											
42.											
43.											
44.											
45.											
46.											
47.											
48.						DATE ______					

Figure 2-15 Tolerance Chart, Partial View Added

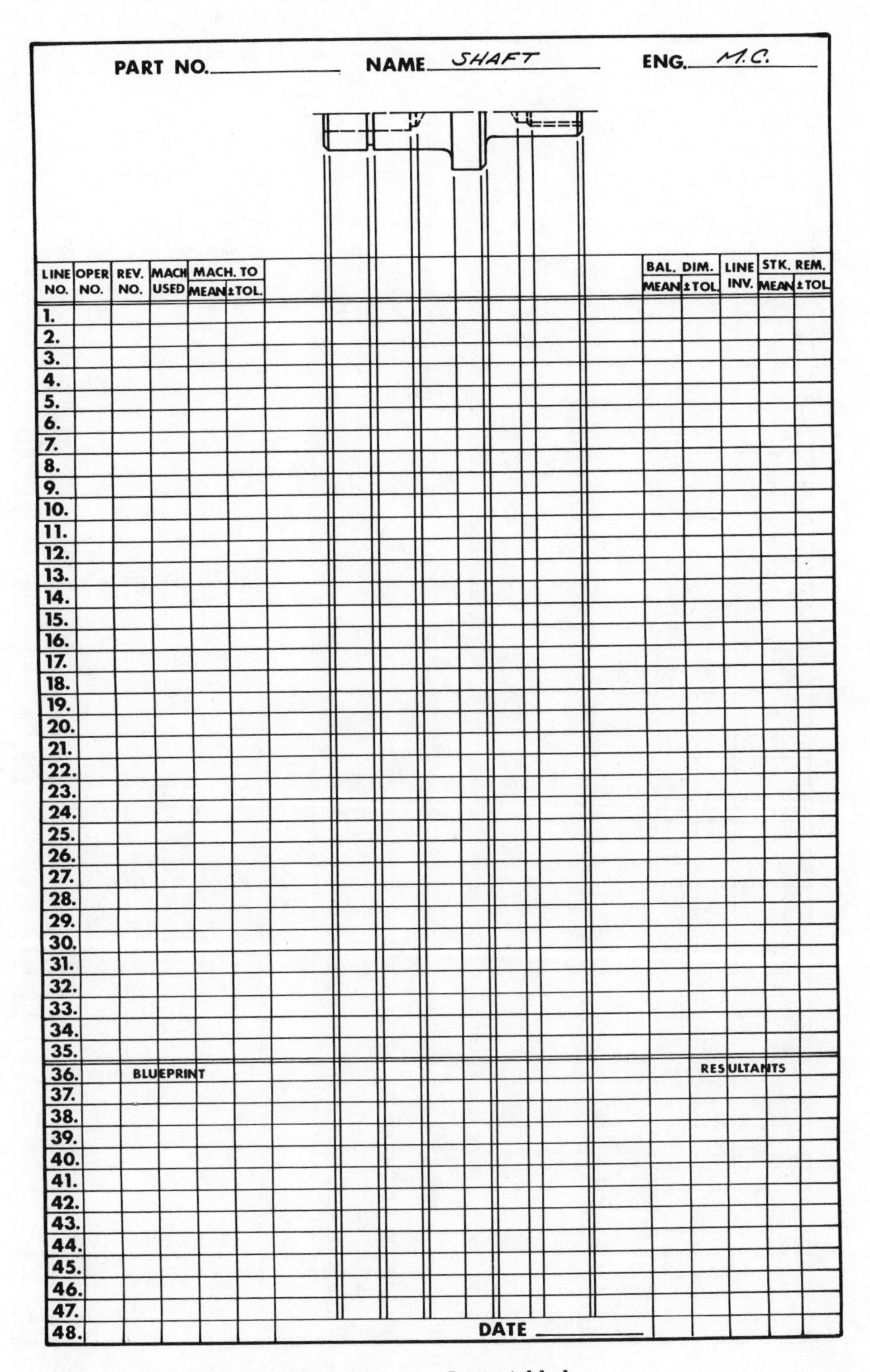

Figure 2-16 Tolerance Chart, Extension Lines Added

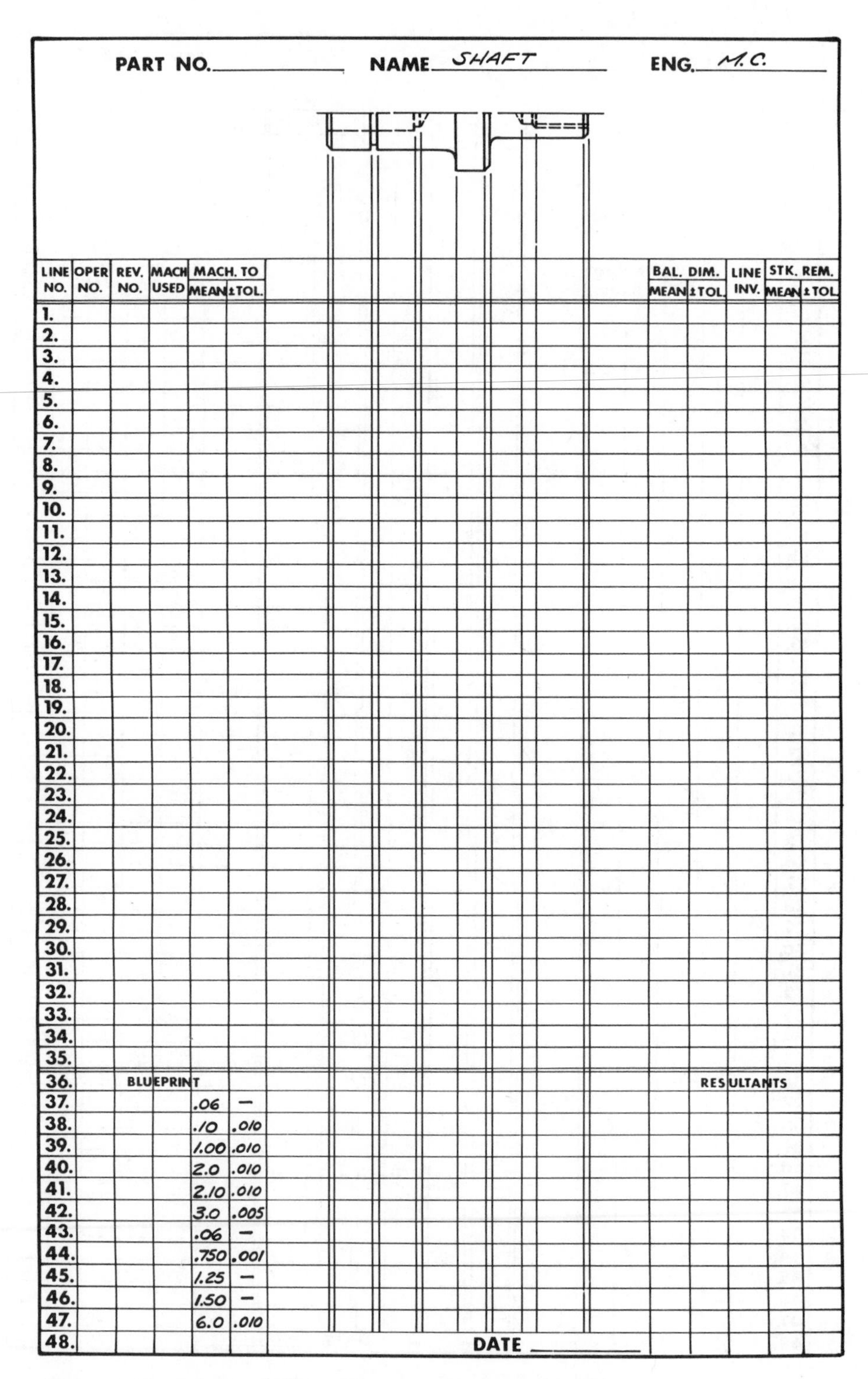

PART NO.________ NAME *SHAFT* ENG. *M.C.*

LINE NO.	OPER NO.	REV. NO.	MACH USED	MACH. TO MEAN	MACH. TO ±TOL.	BAL. DIM. MEAN	BAL. DIM. ±TOL.	LINE INV.	STK. REM. MEAN	STK. REM. ±TOL.
1.										
2.										
3.										
4.										
5.										
6.										
7.										
8.										
9.										
10.										
11.										
12.										
13.										
14.										
15.										
16.										
17.										
18.										
19.										
20.										
21.										
22.										
23.										
24.										
25.										
26.										
27.										
28.										
29.										
30.										
31.										
32.										
33.										
34.										
35.										
36.		BLUEPRINT				RESULTANTS				
37.				.06	—					
38.				.10	.010					
39.				1.00	.010					
40.				2.0	.010					
41.				2.10	.010					
42.				3.0	.005					
43.				.06	—					
44.				.750	.001					
45.				1.25	—					
46.				1.50	—					
47.				6.0	.010					
48.										

DATE ________

Figure 2-17 Tolerance Chart, Blueprint Dimensions Added

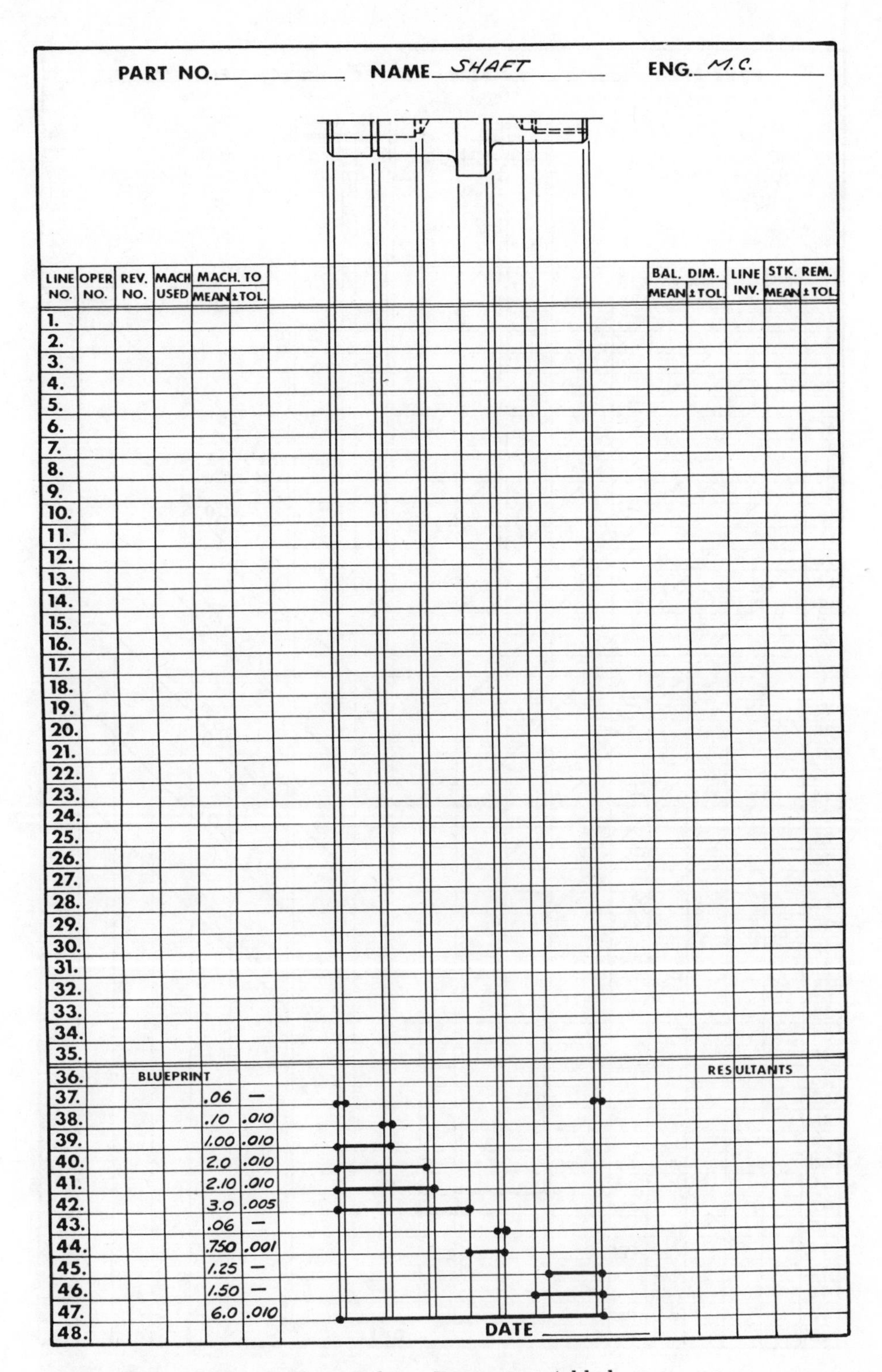

Figure 2-18 Tolerance Chart, Balance Dimensions Added

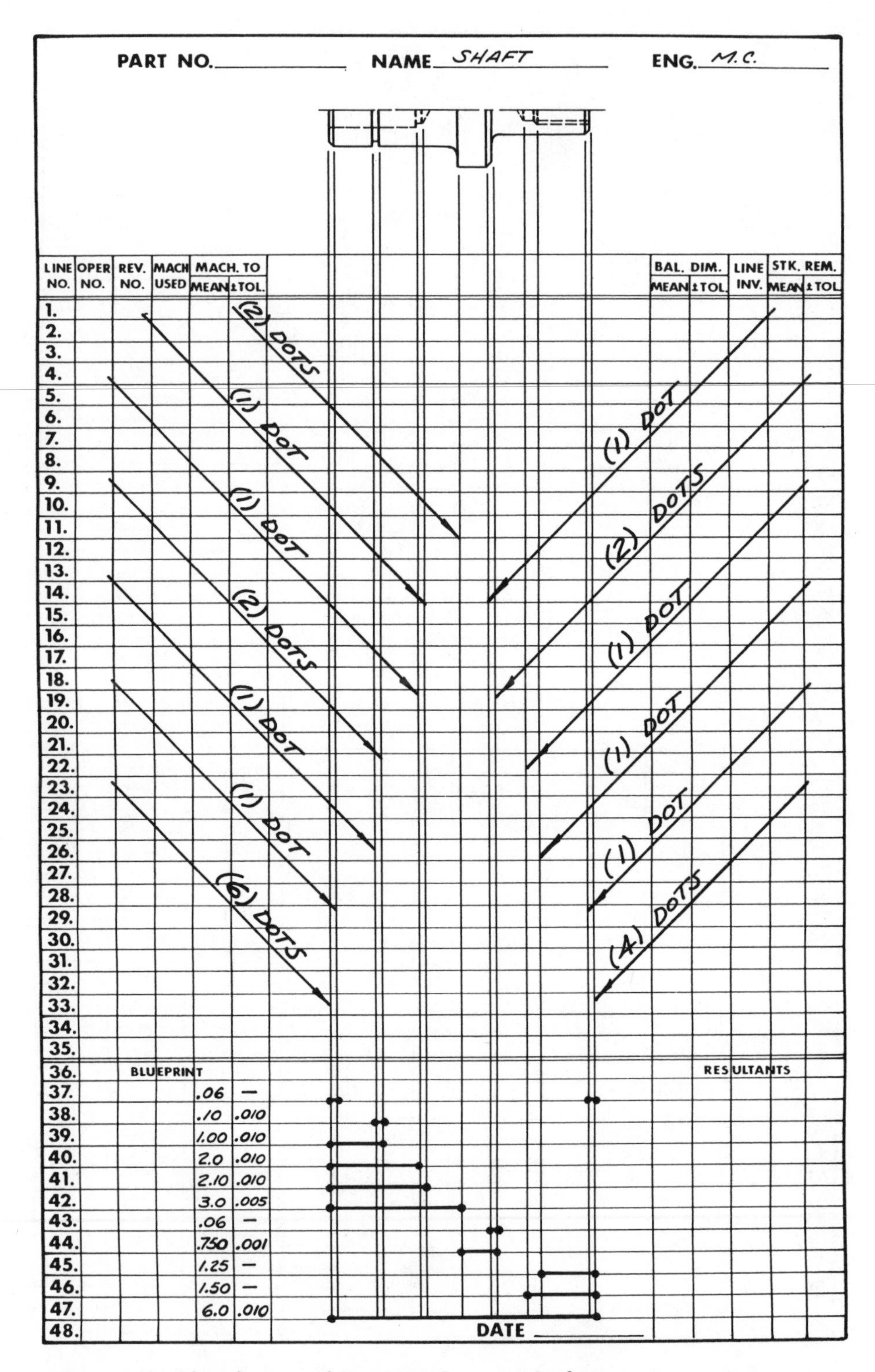

Figure 2-19 Identification of Important Locating Surfaces

This glossary is an alphabetical listing and brief description of many common manufacturing terms. The glossary and the process tables should prove invaluable to an engineer engaged in process selection.

2 ▪ 5 CONCLUSION

Having completed an in-depth analysis of all related part prints, the process planner should have the following items identified for all parts to be made in-house:

1. Raw-material form and structure.
2. Critical dimensions.
3. Clamping- and locating-points.
4. Specifically dictated processes.
5. Other necessary process selections.

This information will now permit work to begin toward the establishment of a formal sequence of operations (called the process plan).

REVIEW QUESTIONS

1. What is a CET and why is it issued?
2. What is the make-or-buy decision all about?
3. List some important make considerations.
4. List some important buy considerations.
5. What happens to parts that are to be purchased?
6. What are the major factors affecting raw-material selection?
7. When part prints are being analyzed, what is the process planner looking for?
8. What is an ECN and what is it used for?
9. What is an ECR and how does it differ from an ECN?
10. What are the major factors affecting process selection?

PROCESS:	APPLICATION INFORMATION		
MACHINING (TRADITIONAL) TYPES	1. TYPICAL APPLICATIONS	2. PREPARATORY PROCESS	3. TYPICAL TOLERANCE AND NOMINAL SURFACE FINISH HELD
1. BORING	TRUING AND SIZING INTERNAL DIAMETER	DRILLING	TOL. ±.001 IN. MICROFINISH - 63
2. BROACHING	RAPIDLY CUTTING INTERNAL AND EXTERNAL GEAR TYPE SHAPES	DRILLING	TOL. ±.002 IN. MICROFINISH - 63
3. COUNTERBORING AND SPOTFACING	CUTTING AN AREA FOR A BOLT HEAD TO SIT	DRILLING	TOL. ±.005 IN. MICROFINISH - 94
4. DRILLING	MAKING HOLE IN SOLID MATERIAL	MILLING, FACING, CORING OR NONE	TOL. $^{+.005}_{-.002}$ IN. MICROFINISH - 125
5. GRINDING	ACCURATELY REMOVING SMALL AMOUNTS OF MATERIAL	MILLING, DRILLING OR BORING	TOL. ±.001 IN. MICROFINISH - 16
6. HOBBING	CUTTING EXTERNAL GEAR TEETH	TURNING	TOL. ±.010 IN. MICROFINISH - 94
7. MILLING	CREATING A DATUM OR LOCATING SURFACE	CASTING, FORGING, COLD OR HOT ROLLING	TOL. ±.010 IN. MICROFINISH - 94
8. REAMING	ACCURATELY SIZING A DRILLED HOLE	DRILLING, DRILLING AND BORING, OR CORING	TOL. ±.001 IN. MICROFINISH - 63
9. SAWING	TURNING BAR STOCK INTO SLUGS	NONE	TOL. ±.050 IN. MICROFINISH - 250
10. SHAPING (GEAR)	CUTTING INTERNAL AND EXTERNAL GEAR TEETH	TURNING	TOL. ±.005 IN. MICROFINISH - 63
11. THREADING	CUTTING INTERNAL AND EXTERNAL THREADS	DRILLING FOR INTERNAL THREADS, TURNING FOR EXTERNAL THREADS	PITCH LINE TOL. ±.003 IN. MICROFINISH - 125
12. TURNING	TRUING AND SIZING EXTERNAL DIAMETERS	CHAMFERING, FACING, OR NONE	TOL. ±.005 IN. MICROFINISH - 94

Figure 2-20 Sample Process Table

APPLICATION INFORMATION			
4. TYPICAL PRODUCTION VOLUME (INCLUDES PART LOAD AND UNLOAD)	5. RELATIVE TOOLING COST	6. DISADVANTAGE TO USE	7. COMMENTS NOMINAL SFPM USING CARBIDE IN MACHINING LOW CARBON STEEL
60/HOUR	LOW	RELATIVELY SLOW OPERATION	350
150/HOUR	HIGH	REQUIRES A SPECIAL MACHINE AND TOOLING	25
60/HOUR	LOW	RELATIVELY SLOW OPERATION AND A WASTE OF MATERIAL	65 - 130
60/HOUR	LOW	RELATIVELY SLOW OPERATION AND A WASTE OF MATERIAL	80
75/HOUR	MEDIUM	EXPENSIVE SECONDARY OPERATION	WORK - 75 WHEEL - 5000
75/HOUR	HIGH	REQUIRES A SPECIAL MACHINE AND GEARING	200 WITH A HSS HOB
40/HOUR	MEDIUM	RELATIVELY SLOW OPERATION AND A WASTE OF MATERIAL	200 - 500
125/HOUR	LOW	SIZES HOLE ONLY. DOES NOT STRAIGHTEN HOLE	ROUGHING - 100 FINISHING - 50
60/HOUR	LOW	SAW CAN ANGLE OFF WHEN CUTTING THICK PARTS	100
50/HOUR	HIGH	REQUIRES A SPECIAL MACHINE AND GEARING	80
100/HOUR	MEDIUM	PROPER TAP AND DIE SELECTION REQUIRES SOME TRIAL AND ERROR	50
60/HOUR	MEDIUM	RELATIVELY SLOW OPERATION AND A WASTE OF MATERIAL	400

Figure 2-20 *(continued)*

CHAPTER 3

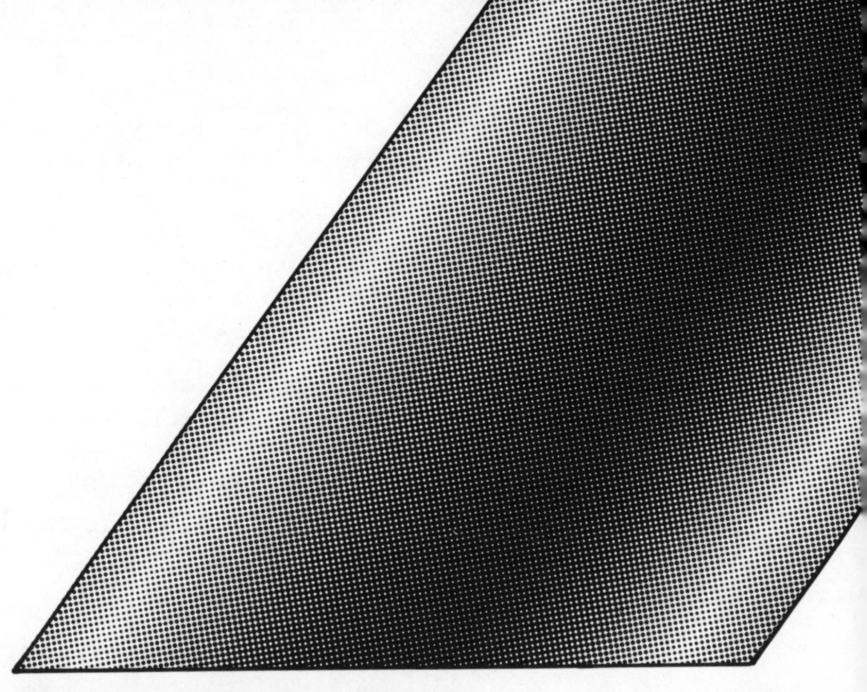

Establishing the Sequence of Processes

3 ▪ 1 INTRODUCTION

After completing the analysis of the part print, the process planner must organize all the information into a logical and economical method of manufacture. Although the required operations may have been previously dictated or selected, the process planner must arrange these operations according to proper sequence. This is now the central concern of the process planner.

By meticulously arranging the processes necessary in the production of any given part, the process planner can achieve one or more of the following economic benefits:

1. Proper use or maximizing of all dimensional tolerances given on the product drawing (see Chapter 6 on tolerance charting).
2. Process elimination.
3. Utilization of in-process operations, which, in effect, results in additional processing without increases in the part's direct labor content.

The following sections of this brief chapter are designed to point out and explain the sequencing considerations used by experienced processing personnel in general manufacturing operations.

3 ▪ 2 TRADITIONAL PROCESS SEQUENCES

Years of process planning will teach the engineer that certain processing sequences or patterns tend to show up time and time again. For example, the tolerance on a given hole might require a certain amount of reaming. Before reaming can begin, however, the hole must be drilled. Even if the hole were cored, the core drilling would have to take place before any reaming could be done. So in this case, the processing pattern of drill-ream is established.

Virtually all production operations tend to fall into such standard or traditional processing sequences. This means that most known production processes are often preceded by other preparatory processes.

To help the process planner with these operational relationships, this book includes a column of information called "Preparatory Operations," which is located on each of the processing tables found in Appendix III (see example in Fig. 3-1). These tables and columns can be used by the planner to work backwards from higher-level processing requirements down to the starting point.

3 ▪ 3 PROCESS ELIMINATION

Remaining competitive in today's manufacturing arena has become a difficult task. Customers continue to demand higher quality, lower prices, and more individual options. Because of high interest rates many investors have placed their capital in safer and more lucrative ventures. In addition, many of the world's nations are gaining manufacturing expertise here-to-fore unseen by the West. These factors have forced all manufacturers to rethink the way they do business.

For years companies have developed hasty process plans to expedite entry into various markets. This practice was based on the idea of getting into production first and cleaning up the process plan later. Of course the cleanup effort usually resulted in cost savings through better manufacturing methods, elimination of wasteful procedures, and process combination. However, this plan of doing it right the second time is very costly in terms of capital equipment, durable tooling, direct labor, and administrative overhead. In fact, today's competitive manufacturing atmosphere has made this philosophy financially obsolete. Therefore, the modern process planner must do it right the first time. This means the planner must analyze, from the outset, all required processes for possible combinations or eliminations. The following examples illustrate how this might happen.

EXAMPLE 1

The die-cast part shown in Figure 3-2 was originally processed according to the routing shown as Figure 3-3. Later, roll-forming taps were substituted for the cut taps first used. The result, as shown in Figure 3-4, was the elimination of part of operation 20 (blow out chips) and an increase in pieces per hour, which means improved productivity. In this case, one minor change simultaneously increased quality and lowered cost.

EXAMPLE 2

The small spacer block shown in Figure 3-5 was originally processed according to the routing shown in Figure 3-6. Later, the process was changed as shown in Figure 3-7. This second routing eliminated the deburring operation and its cost. Surface *B* was milled after the hole was drilled, so the burr created by the drill breakthrough was cut off in the milling of operation 30. Although a small burr was wiped into the hole by the milling operation, the hole was for bolt clearance only; the clearance was functionally unaffected by the milling burr.

In this example, as with the first, subtle and inexpensive changes eliminated processing labor and its related equipment costs. This kind of thinking cannot be an afterthought; it must be a conscious and deliberate effort that takes place as the original process plan is developed.

PROCESS:	APPLICATION INFORMATION		
MACHINING (TRADITIONAL) TYPES	1. TYPICAL APPLICATIONS	2. PREPARATORY PROCESS	3. TYPICAL TOLERANCE AND NOMINAL SURFACE FINISH HELD
1. BORING	TRUING AND SIZING INTERNAL DIAMETER	DRILLING	TOL. ±.001 IN. MICROFINISH - 63
2. BROACHING	RAPIDLY CUTTING INTERNAL AND EXTERNAL GEAR TYPE SHAPES	DRILLING	TOL. ±.002 IN. MICROFINISH - 63
3. COUNTERBORING AND SPOTFACING	CUTTING AN AREA FOR A BOLT HEAD TO SIT	DRILLING	TOL. ±.005 IN. MICROFINISH - 94
4. DRILLING	MAKING HOLE IN SOLID MATERIAL	MILLING, FACING, CORING OR NONE	TOL. $^{+.005}_{-.002}$ IN. MICROFINISH - 125
5. GRINDING	ACCURATELY REMOVING SMALL AMOUNTS OF MATERIAL	MILLING, DRILLING OR BORING	TOL. ±.001 IN. MICROFINISH - 16
6. HOBBING	CUTTING EXTERNAL GEAR TEETH	TURNING	TOL. ±.010 IN. MICROFINISH - 94
7. MILLING	CREATING A DATUM OR LOCATING SURFACE	CASTING, FORGING, COLD OR HOT ROLLING	TOL. ±.010 IN. MICROFINISH - 94
8. REAMING	ACCURATELY SIZING A DRILLED HOLE	DRILLING, DRILLING AND BORING, OR CORING	TOL. ±.001 IN. MICROFINISH - 63
9. SAWING	TURNING BAR STOCK INTO SLUGS	NONE	TOL. ±.050 IN. MICROFINISH - 250
10. SHAPING (GEAR)	CUTTING INTERNAL AND EXTERNAL GEAR TEETH	TURNING	TOL. ±.005 IN. MICROFINISH - 63
11. THREADING	CUTTING INTERNAL AND EXTERNAL THREADS	DRILLING FOR INTERNAL THREADS, TURNING FOR EXTERNAL THREADS	PITCH LINE TOL. ±.003 IN. MICROFINISH - 125
12. TURNING	TRUING AND SIZING EXTERNAL DIAMETERS	CHAMFERING, FACING, OR NONE	TOL. ±.005 IN. MICROFINISH - 94

Figure 3-1 Preparatory Operations as Shown On Process Tables

APPLICATION INFORMATION			
4. TYPICAL PRODUCTION VOLUME (INCLUDES PART LOAD AND UNLOAD)	5. RELATIVE TOOLING COST	6. DISADVANTAGE TO USE	7. COMMENTS NOMINAL SFPM USING CARBIDE IN MACHINING LOW CARBON STEEL
60/HOUR	LOW	RELATIVELY SLOW OPERATION	350
150/HOUR	HIGH	REQUIRES A SPECIAL MACHINE AND TOOLING	25
60/HOUR	LOW	RELATIVELY SLOW OPERATION AND A WASTE OF MATERIAL	65 - 130
60/HOUR	LOW	RELATIVELY SLOW OPERATION AND A WASTE OF MATERIAL	80
75/HOUR	MEDIUM	EXPENSIVE SECONDARY OPERATION	WORK - 75 WHEEL - 5000
75/HOUR	HIGH	REQUIRES A SPECIAL MACHINE AND GEARING	200 WITH A HSS HOB
40/HOUR	MEDIUM	RELATIVELY SLOW OPERATION AND A WASTE OF MATERIAL	200 - 500
125/HOUR	LOW	SIZES HOLE ONLY. DOES NOT STRAIGHTEN HOLE	ROUGHING - 100 FINISHING - 50
60/HOUR	LOW	SAW CAN ANGLE OFF WHEN CUTTING THICK PARTS	100
50/HOUR	HIGH	REQUIRES A SPECIAL MACHINE AND GEARING	80
100/HOUR	MEDIUM	PROPER TAP AND DIE SELECTION REQUIRES SOME TRIAL AND ERROR	50
60/HOUR	MEDIUM	RELATIVELY SLOW OPERATION AND A WASTE OF MATERIAL	400

Figure 3-1 *(continued)*

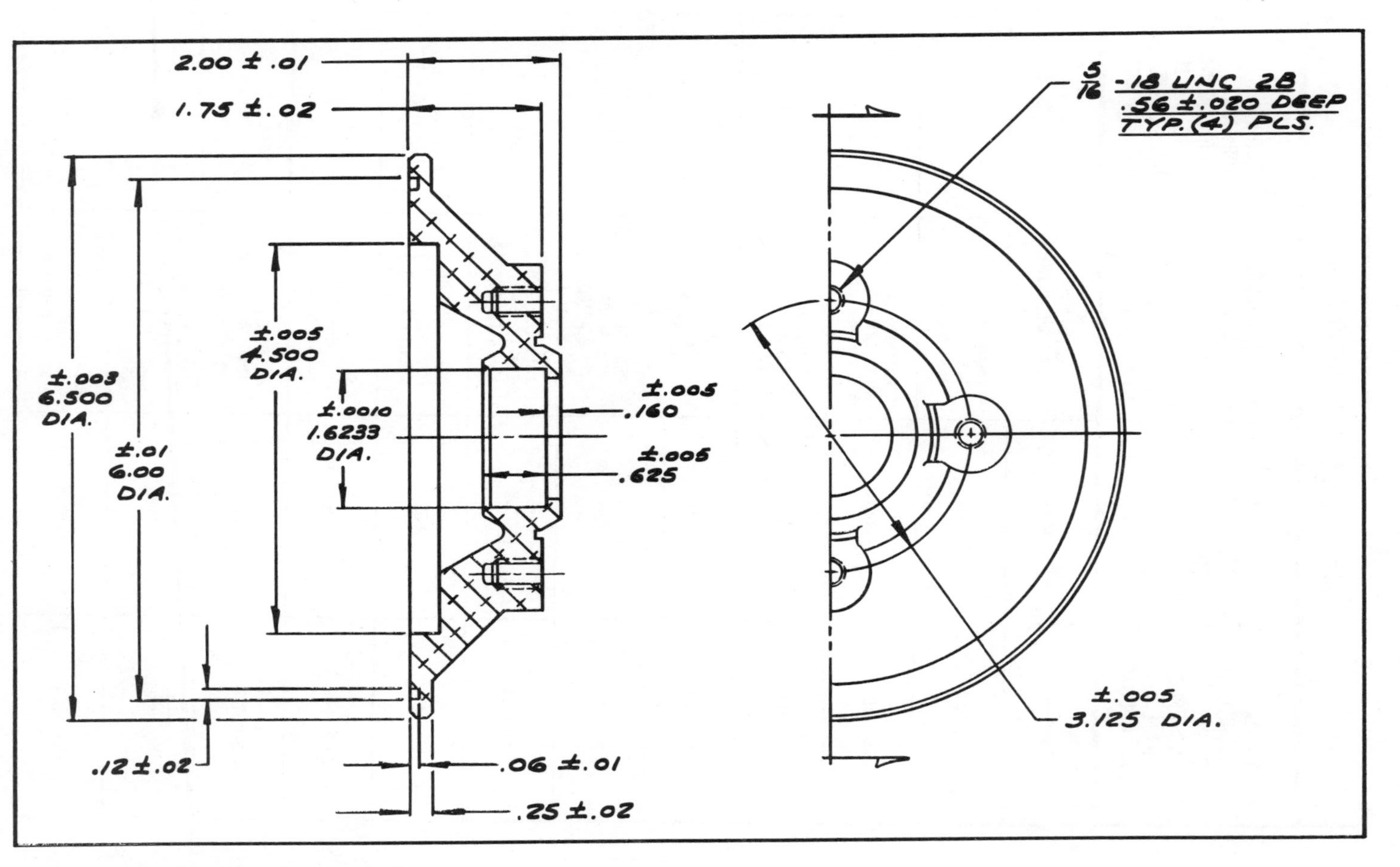

Figure 3-2 Sample Part, Viscous Body

NAME VISCOUS BODY DATE 1-12-87 PART NO. 30143
ASSY. NO. 15587 CUST. G.M.
MATL. ALU. SHEET NO. 1

Dept	Description	Sheet	Oper	Hourly Production	Rate	Labor	Set-Up Hours
15-2	MACHINE BODY & POCKET	2	10	150	12.01	.08	4
15-2	TAP (4) HOLES & BLOW OUT CHIPS	3	20	150	12.01	.08	2
11-4	WASH	4	30	1000	11.27	.0113	—
15-7	IMPREGNATE	5	40	2000	11.27	.0056	—
21-2	CHECK BALANCE	6	50	200	10.81	.054	1

Date	Change	By	Date	Change	By	Date	Change	By
1-12-87	RELEASED FOR PRODUCTION	M.C.						

Figure 3-3 Sample Routing

NAME VISCOUS BODY DATE 1-12-87 PART NO. 30143
ASSY. NO. 15587 CUST. G.M.
MATL. ALU. SHEET NO. 1

Dept	Description	Sheet	Oper	Hourly Production	Rate	Labor	Set-Up Hours
15-2	MACHINE BODY & POCKET	2	10	150	12.01	.08	4
15-2	TAP (4) HOLES	3	20	250	12.01	.048	2
11-4	WASH	4	30	1000	11.27	.0113	—
15-7	IMPREGNATE	5	40	2000	11.27	.0056	—
21-2	CHECK BALANCE	6	50	200	10.81	.054	1

Date	Change	By	Date	Change	By	Date	Change	By
1-12-87	RELEASED FOR PRODUCTION	M.C.						
3-4-87	OPER. 20 WAS 150/HR.	M.C.						

Figure 3-4 Sample Routing Revised

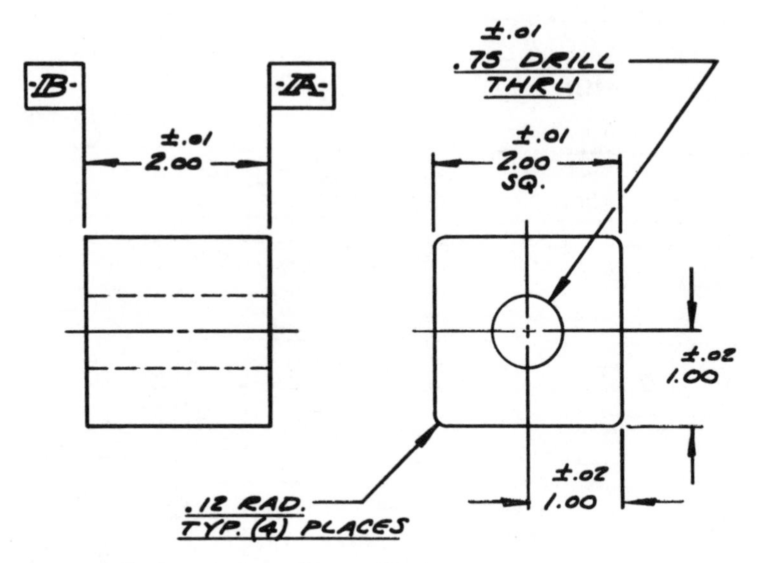

Figure 3-5 Sample Part, Spacer Block

NAME SPACER BLK. **DATE** 1-14-87 **PART NO.** 94275
ASSY. NO. 12461 **CUST.** ABC INC.
MATL. SAE 1018 **SHEET NO.** 1

Dept	Description	Sheet	Oper	Hourly Production	Rate	Labor	Set-Up Hours
11-5	MILL SURFACE "A"	2	10	250	12.56	.0502	1
11-5	MILL SURFACE "B"	3	20	250	12.56	.0502	1
10-2	DRILL THRU HOLE	4	30	125	10.75	.086	1
09-3	HAND DEBURR HOLE	5	40	275	9.81	.0356	—
	@ SURFACE "B"						

Date	Change	By	Date	Change	By	Date	Change	By
1-14-87	RELEASED FOR							
	PRODUCTION	MC						

Figure 3-6 Sample, Example 2

NAME SPACER BLK DATE 1-14-87 PART NO. 94275

ASSY. NO. 12461 CUST. ABC INC.

MATL. SAE 1018 SHEET NO. 1

Dept	Description	Sheet	Oper	Hourly Production	Rate	Labor	Set-Up Hours
11-5	MILL SURFACE "A"	2	10	250	12.56	.0502	1
10-2	DRILL THRU HOLE	3	20	125	10.75	.086	1
11-5	MILL SURFACE "B"	4	30	250	12.56	.0502	1

Date	Change	By	Date	Change	By	Date	Change	By
1-14-87	RELEASED FOR		4-11-87	REMOVED OPER				
	PRODUCTION	M.C.		#40 DEBURRING	M.C.			
4-11-87	OPER 20 WAS 30							
	AND 30 WAS 20	M.C.						

Figure 3-7 Revised Routing, Example 2

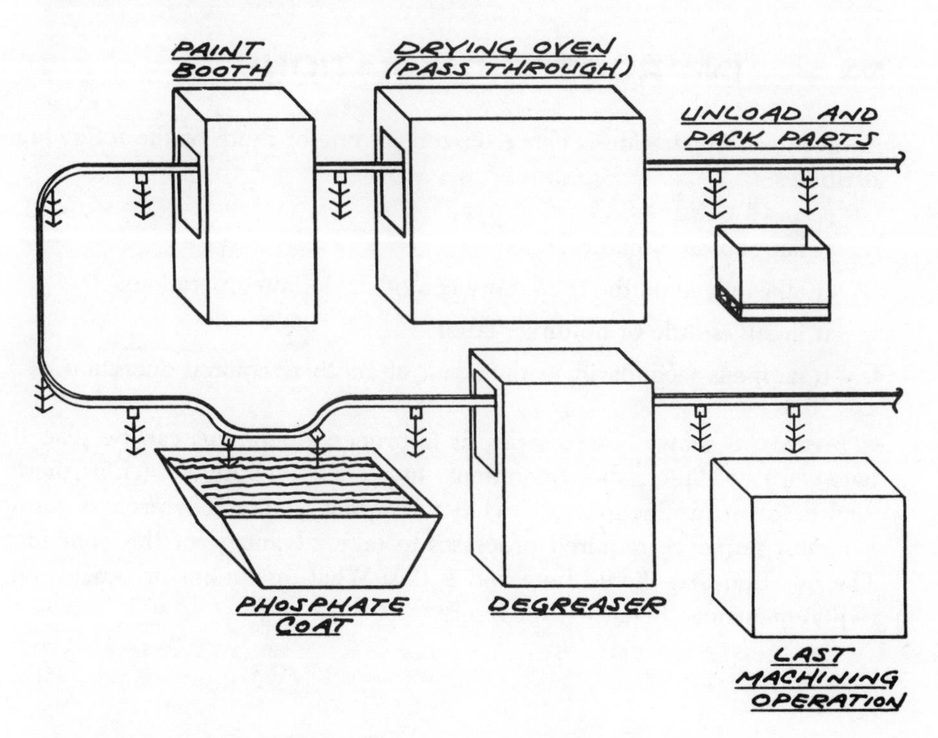

Figure 3-8 In-Process Operations

NAME ANY PART DATE 1-14-87 PART NO. ~

ASSY. NO. ~ CUST. ~

MATL. ~ SHEET NO. 1

Dept	Description	Sheet	Oper	Hourly Production	Rate	Labor	Set-Up Hours
12-5	LAST MACHINING OPERATION	2	10A	150	12.56	.0837	1
12-5	— DEGREASE	3	10B	—	—	—	—
12-5	— PHOSPHATE COAT	4	10C	—	—	—	—
12-5	— PAINT	5	10D	—	—	—	—
12-5	— DRY	6	10E	—	—	—	—
12-5	UNLOAD AND PACK	7	20	150	10.21	.0681	—

Date	Change	By	Date	Change	By	Date	Change	By
1-14-87	RELEASED FOR							
	PRODUCTION	MC						

Figure 3-9 Routing of In-Process Operations

3 ▪ 4 IN-PROCESS OPERATIONS

An in-process operation is characterized by one or more of the following attributes:

1. It happens automatically.
2. It takes place on the route between other definite operations.
3. It involves little or no direct labor.
4. It happens secondarily as the result of another required operation.

Figure 3-8 shows how a series of in-process operations can be placed between two direct labor operations. In-process operations such as these tend to lower production costs while increasing quality. The process planner must sequence required processes to take advantage of this concept. The question that should be asked is this: What operations or processing requirements can be gotten for free?

3 ▪ 5 CONCLUSION

At this point, the process planner should know what processes are required to produce a given part and how those processes can best be sequenced. A formal routing (see Fig. 3-9) should be near completion. The planner is now prepared to address a variety of economic questions, which result from this tentative and workable process plan.

REVIEW QUESTIONS

1. What are the major benefits realized as a result of proper process sequencing?
2. What is meant by the phrase "traditional or standard process sequence"?
3. Why has it been difficult for U.S. manufacturing companies to remain competitive in world and domestic markets?
4. How can the process planner help to make his or her company more competitive?
5. What are the characteristics of "in-process operations"?

CHAPTER 4

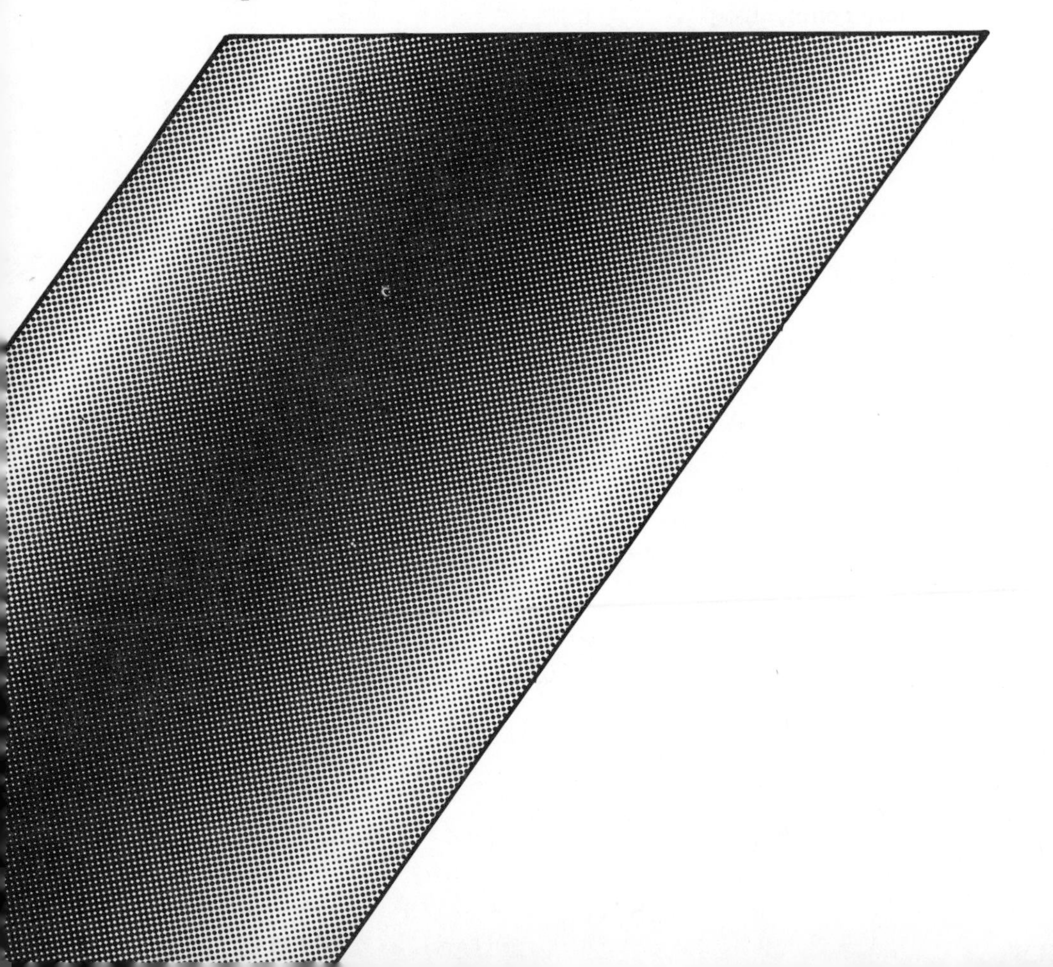

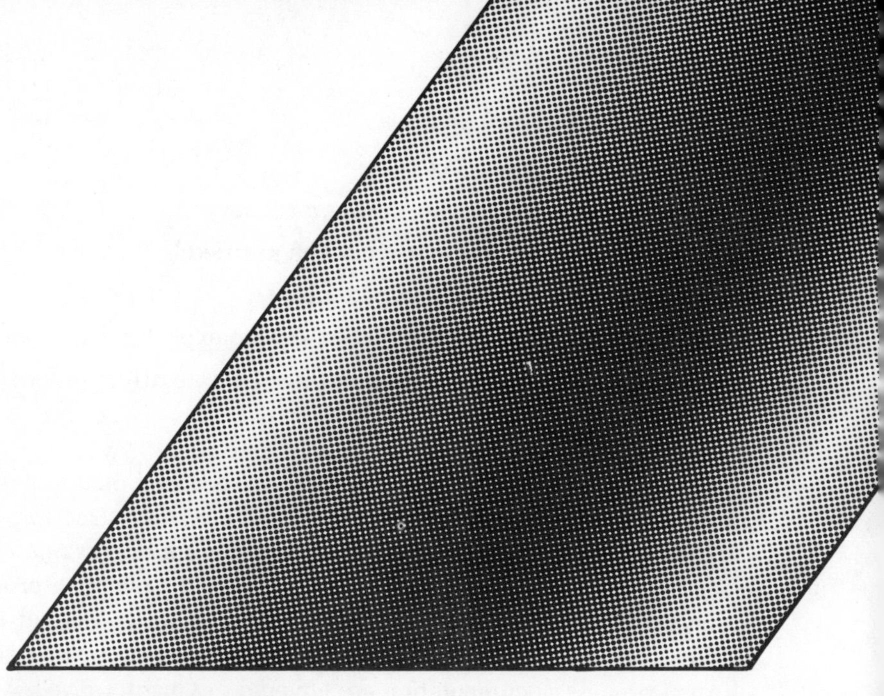

Economic Processing Considerations

4 ▪ 1 INTRODUCTION

Chapter 2 discussed how the process planner analyzes a part by using the graphic representation on the print. That chapter also pointed out that the process planner must identify the basic processes required to produce the part. Chapter 3 found the engineer trying to logically sequence these basic processes. Now, several questions affecting the profitability of each operation must be answered.

1. What machine is required?
2. Will the machine be capable?
3. Is there capacity on an existing machine?
4. Can the cost of a new machine be justified?
5. How should the machine be tooled?
6. What kind of hourly production can be expected from one machine?
7. How does the plant's labor-and-burden rate affect other required decisions?

Answers to many of these questions are recorded on a routing-like form sometimes referred to as an *operation-and-cost sheet* (see Fig. 4-1). Other related bits of information might be placed on a *cost-estimate transmittal* (see Fig. 4-2). All back-up data in this phase of the planning process—items such as sketches, notes, charts, figures, and so forth—should be saved in an appropriately labeled file folder. It should be noted that all phases of formal process documentation are covered in Chapter 6.

NAME ________ DATE ________ PART NO. ________

ASSY. NO. ________ CUST. ________

MATL. ________ SHEET NO. ________

Dept	Description	Sheet	Oper	Hourly Production	Rate	Labor	Set-Up Hours

Date	Change	By	Date	Change	By	Date	Change	By

Figure 4-1 Operation-and-Cost Sheet

COST
ESTIMATE
TRANSMITTAL

CET NO.______
DATE OF ISSUE______
DATE REQUIRED______

DATA REQUIRED

☐ FIRM PROCESS ☐ PRELIMINARY PROCESS
☐ FIRM MATERIAL ☐ PRELIMINARY MATERIAL
☐ ADD-DELETE ☐ OTHER
☐ NEW ☐ REPEAT BUSINESS

DATA KNOWN

☐ CUST. EST. ☐ SALES EST.
LOT SIZE______ PACK SIZE______
PER MO.______ ANNUAL______
PROD. TARGET DATE______
REPLACES PART NO.______
DRAWINGS AVAILABLE YES NO

SALES COMMENTS (SALES ENGINEERING)

MANUFACTURING COMMENTS (PROCESS ENGINEERING)

LEAD TIME FROM FORMAL PROGRAM RELEASE TO:
1) PRODUCTION READY
2) DELIVERY OR COMPLETION IN HOUSE

CAPITAL EQUIPMENT ______WKS. $______
VENDOR TOOLING ______ ______
IN PLANT TOOLING ______ ______
GAGES & Q.C. ______ ______
SETUP ______ ______
TOTAL ______WKS. $______

CC: GENERAL MGR. MFG. MGR. MKTG. MGR. ADV. ENG MGR. CH. PROD. ENG. CH. MFG. ENG.
SALES MGR. MATERIALS MGR. PRODUCTION MGR. PURCHASING MGR. CONTROLLER Q.C. ENG.

POSITION	SIGNATURE	DATE
PREPARED BY		
PROC. ENG.		
Q.C. ENG.		
CH. MFG. ENG.		
MFG. MGR.		

Figure 4-2 The Cost-Estimate Transmittal

The balance of this chapter is devoted to addressing questions, such as those previously listed, which are likely to face the process planner at this point.

4 ▪ 2 MACHINE SELECTION

Because there are a variety of complicating variables, selecting a machine capable of accommodating the individual processing needs of each operation can be a difficult task.

First, the process planner must choose the best *machine tool* for the job. Machine tools are also referred to as *capital equipment,* and there is often more than one type or style available. The best choice will be contingent upon other economic factors such as the following:

1. tooling cost
2. set-up cost
3. frequency of changeover
4. versatility requirements (that is, can it easily run dissimilar parts?)
5. production rate
6. basic machine cost
7. ease of automation
8. direct labor costs within the plant

To help the engineer research and organize these variables, this book contains machine-tool classification tables that cover these factors (see Appendix IV).

Break-Even Charts

Machine selection data can also be quantified and graphically displayed on a *break-even chart* (see Fig. 4-3). It is one popular and convenient tool used by process planners in comparing two or more sets of economic data. In this case, it compares two machines to determine which is more economical.

In Figure 4-3, note how the fixed cost of each machine is plotted. This fixed cost includes the cost of the capital equipment, plus any related durable tooling cost. On top of each fixed-cost line, a related variable cost line is drawn. It represents the cost per piece including, but not limited to, the following items: direct labor, direct material, perishable tools, and set-up costs.

In this example (Fig. 4-3), the cost of machine 1 is $20,000, and the cost per piece is 60 cents. The cost of machine 2 is $40,000, and the cost per piece is 10 cents. The chart indicates that machine 1 is more economical for total production quantities of 38,000 pieces or less. The chart also indicates that machine 2 is more economical for total production quantities in excess of 38,000.

Payback Comparison

Another tool used by process planners to justify capital equipment expenditures is called the *payback comparison* method. This method requires

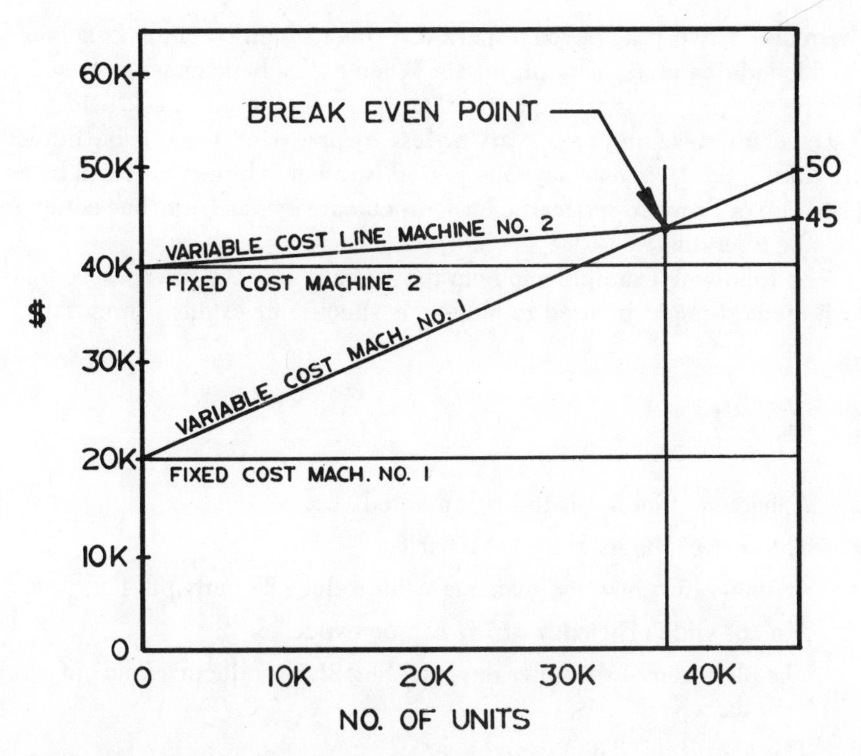

Figure 4-3 The Break-Even Chart

that the original investment be recovered in a specified length of time (expressed in years) from either the profits or savings generated by the capital expenditure.

The formula used for calculating the time an investment takes to pay for itself is as follows:

$$\frac{\text{Investment Cost in Dollars}}{\text{Annual Profits} - \text{Annual Operating Expenses}} = \begin{array}{c}\text{Payback Period}\\ \text{in Years}\end{array}$$

or

$$\frac{\text{Investment Cost in Dollars}}{\text{Net Annual Savings}} = \begin{array}{c}\text{Payback Period}\\ \text{in Years}\end{array}$$

The dollar figures used in applying these formulas are direct, meaning that the effects of inflation, compound interest, depreciation, and salvage value are not considered.

More elaborate methods of financial analysis do exist and are sometimes used. However, a majority of manufacturing companies have selected the payback comparison method for use in justifying capital purchases. Furthermore, a two-year payback period seems to be a common benchmark used in judging investment proposals. When such a benchmark is the guide, investment proposals taking more than two years to pay back tend to be rejected; those taking two years or less to pay back tend to be funded. Of course, this two-year payback period is rather arbitrary and will be adjusted up or down to reflect the business climate in which any one company may be operating.

The following example will help to illustrate how the payback comparison method might be used to judge one specific investment proposal.

FACTS

1. A machine tool needs to be purchased.
2. The cost of the machine is $100,000.
3. At 100% efficiency, the machine will produce 20 parts per hour.
4. A plant-wide efficiency of 75% can be expected.
5. The plant runs 3 shifts per day, yielding 21.5 productive hours of work each day.
6. The plant runs 250 days per year.
7. The sales department would like to charge 30 cents per part as profit on all parts manufactured.
8. A long-term contract will be awarded.
9. The company likes a two-year payback on all capital investments.

FIGURES

		20.00	pieces/hour @ 100%
Plant efficiency	×	0.75	
		15.00	pieces/hour (net output)
Productive hours/day	×	21.50	
		322.50	pieces/day
Working days/year	×	250.00	
		80,625.00	pieces/day
Profit/piece in $	×	0.30	
		24,187.50	annual profits

FORMULA

$$\frac{\text{Investment Cost in Dollars}}{\text{Annual Profits} - \text{Annual Operating Costs}} = \text{Payback Period in Years}$$

$$\frac{\$100{,}000}{\$24{,}187.50 - 0} = 4.13 \text{ Years Payback}$$

A 4.13-year payback in this case violates the company's desire for a 2-year payback and would therefore be rejected. However, by rearranging the payback equation, the engineer can figure out what must be charged as a profit to make this example a good investment.

FORMULA

$$\frac{\text{Investment Cost in Dollars}}{\text{Required Payback in Years}} = \text{Required Annual Profit}$$

EXAMPLE

$$\frac{\$100{,}000 \text{ (investment)}}{2 \text{ (years required for payback)}} = \$50{,}000 \text{ (in required annual profit)}$$

$$\frac{\$50{,}000 \text{ Annual Profit}}{80{,}625 \text{ Parts/Year}} = \$0.62/\text{part}$$

Profit that must be charged

Existing Machinery

Whenever machines are being selected for use in a process plan, a plant's existing machinery must be considered. In this context, the term *existing machinery* refers to those pieces of capital equipment presently owned by the process planner's company. The machinery could either be in storage or in production turning out parts other than those being considered in the process plan.

If existing machinery can be used, great financial benefits will be realized due to the following factors:

1. The equipment was previously justified and paid for with prior or existing business.
2. The equipment's true capability and potential for productivity is known.
3. The equipment is in-house; therefore, there are no waiting or delivery problems.
4. People in the plant are already familiar with the equipment. The need for training is thereby minimized, and the opportunity for immediate productivity is enhanced.

However, existing machinery is not without its problems. For example, equipment presently in production is used a given number of hours each month. These hours of use are expressed in terms of percent utilization. A machine used one-half of the month's available straight-time hours would be considered at 50 percent utilization. If the planner's production needs can be accommodated with the remaining or unused hours on the machine, it would be a good candidate for selection. Furthermore, when a variety of parts are run on a single machine, production scheduling becomes difficult. This is especially true when one considers how numerous changeovers and the corresponding downtime cut into the hours available to run production.

The age and maintenance condition of an existing machine must also be considered. If a machine has had a long history of maintenance trouble, it should not be taxed with the additional wear and tear of more production. When there is doubt about the condition of an existing machine tool, the maintenance supervisor should be consulted, and the number of work-order hours that have been charged to that machine over the preceding two years or so should be checked.

Also, whenever a machine tool is in storage or appears on the corporation's surplus equipment list, the process planner should be on guard. He or she should find out why the machine has been taken out of service before using it as the foundation in a new process plan.

Machine Capability

Before a process engineer can confidently select an existing machine or process or approve payment for new capital equipment, he or she must be assured of the machine's ability to repeat and hold necessary tolerances. In industrial terms, checking this ability is called taking a *machine-* or *process-capability study*.

Such a check requires the use of statistics and control charts. These items allow the engineer to draw statistically sound conclusions from small amounts of data.

A process-capability study is used to measure those variations in the process that result from common or normal causes. This definition means that a broken or dull tool would not be considered a common cause. In effect, the process-capability study is only accurate when everything is set up and running correctly (within statistical control).

AVERAGE-AND-RANGE CHART

To make sure the process is running within statistical control, an average-and-range chart must be constructed. $\overline{X}$ is the symbol for average, and R indicates range. The characteristic being studied in this example (see Fig. 4-4) is a bored bearing pocket that has a print specification of 1.6233 in. ± .0010 in. for a total tolerance of .002 in.

DATA COLLECTION

First, 50 consecutive parts should be run and measured. These parts should be run over a short period of time, and every 5 parts run should be placed in one subgroup. When finished, the engineer should have 10 subgroups of 5 parts each.

The measurement data for each part is then placed on the control chart (see Fig. 4-5). Next, the sum of the measurement of each 5-part subgroup is divided by 5, which yields an average or $\overline{X}$. Then the range R in measurement found in each subgroup must be calculated and recorded. Range equals the largest dimension minus the smallest dimension.

CONTROL-CHART CONSTRUCTION

The average ($\overline{X}$) and range (R) for each of the subgroups is then plotted on the average-and-range chart (see Fig. 4-6). An average of the averages, referred to as the grand average ($\overline{\overline{X}}$), and the average of the ranges ($\overline{R}$) are calculated using the data from each subgroup:

$$\overline{\overline{X}} = \frac{(\overline{X}_1 + \overline{X}_2 + \overline{X}_3 \ldots\ldots \overline{X}_{10})}{10}$$

$$\overline{R} = \frac{(R_1 + R_2 + R_3 \ldots\ldots R_{10})}{10}$$

These calculated values are then plotted on the average and range charts respectively (see Fig. 4-7).

Upper and lower control limits for both the average and range charts can now be calculated using the following formulas:

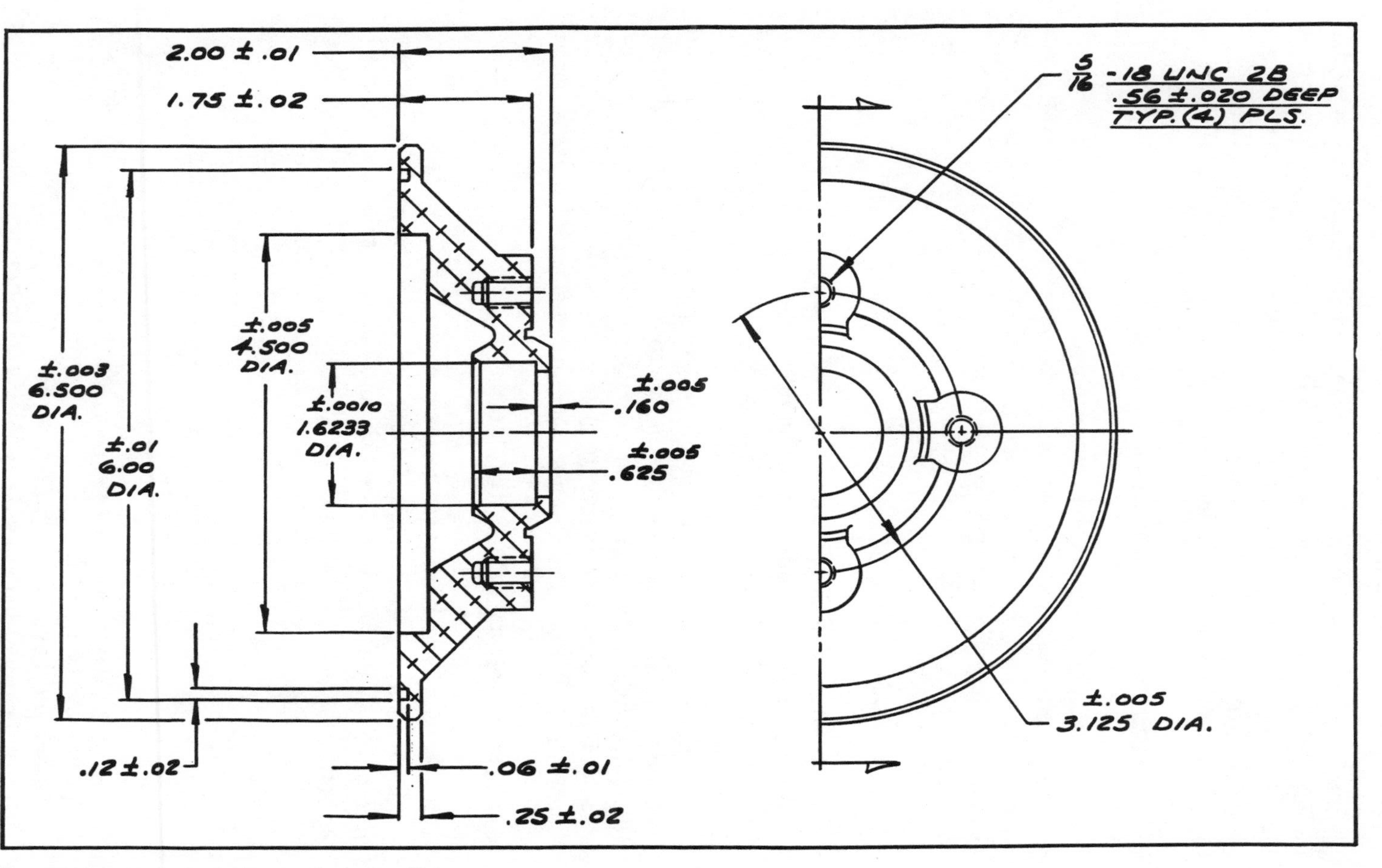

Figure 4-4 Sample Part, Machine Capability

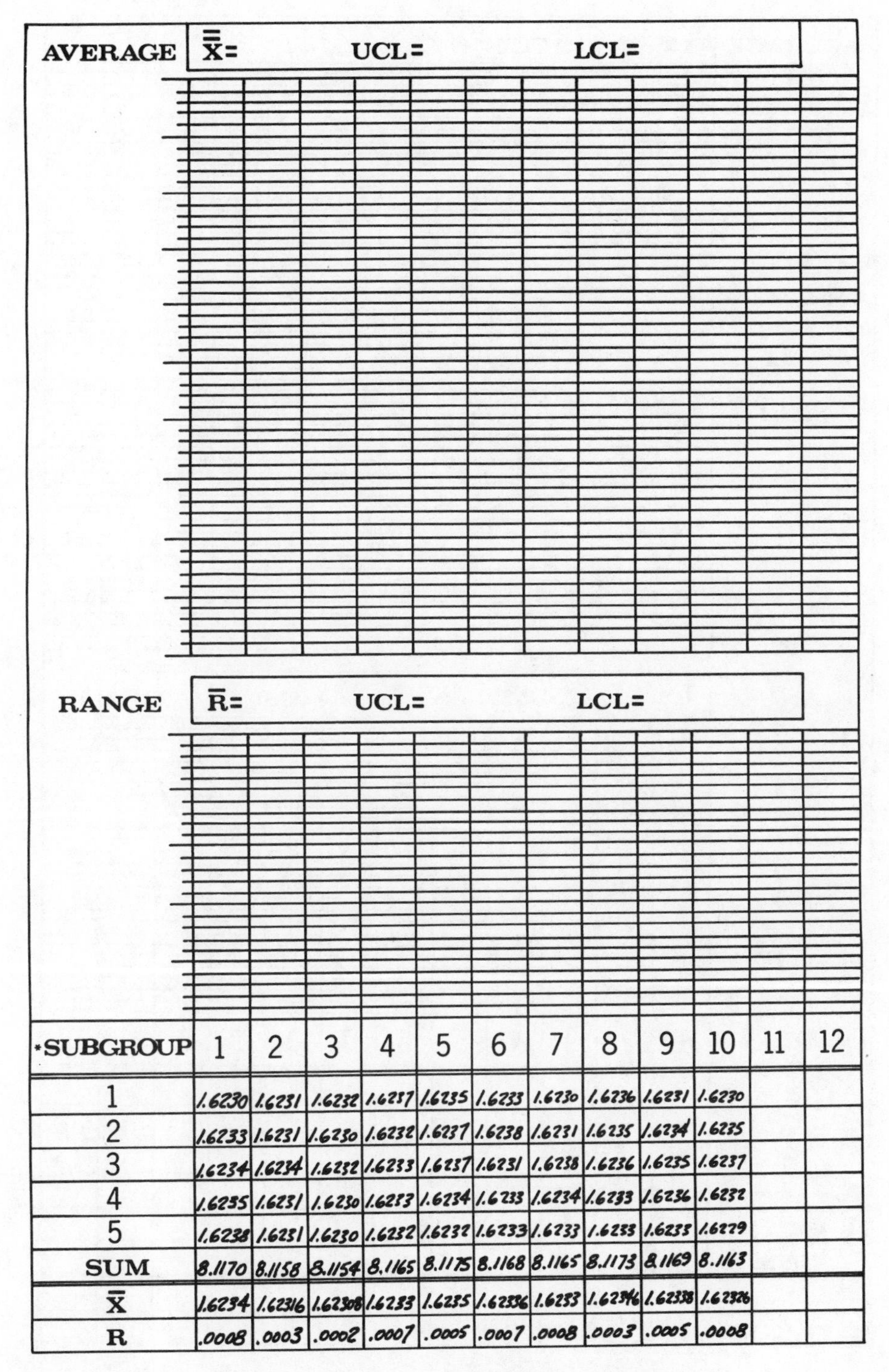

*SUBGROUP	1	2	3	4	5	6	7	8	9	10	11	12
1	1.6230	1.6231	1.6232	1.6237	1.6235	1.6233	1.6230	1.6236	1.6231	1.6230		
2	1.6233	1.6231	1.6230	1.6232	1.6237	1.6238	1.6231	1.6235	1.6234	1.6235		
3	1.6234	1.6234	1.6232	1.6233	1.6237	1.6231	1.6238	1.6236	1.6235	1.6237		
4	1.6235	1.6231	1.6230	1.6233	1.6234	1.6233	1.6234	1.6233	1.6236	1.6232		
5	1.6238	1.6231	1.6230	1.6232	1.6232	1.6233	1.6233	1.6233	1.6233	1.6229		
SUM	8.1170	8.1158	8.1154	8.1165	8.1175	8.1168	8.1165	8.1173	8.1169	8.1163		
$\bar{X}$	1.6234	1.62316	1.62308	1.6233	1.6235	1.62336	1.6233	1.62346	1.62338	1.62326		
R	.0008	.0003	.0002	.0007	.0005	.0007	.0008	.0003	.0005	.0008		

Figure 4-5 Recording the Data for Average-and-Range Charting

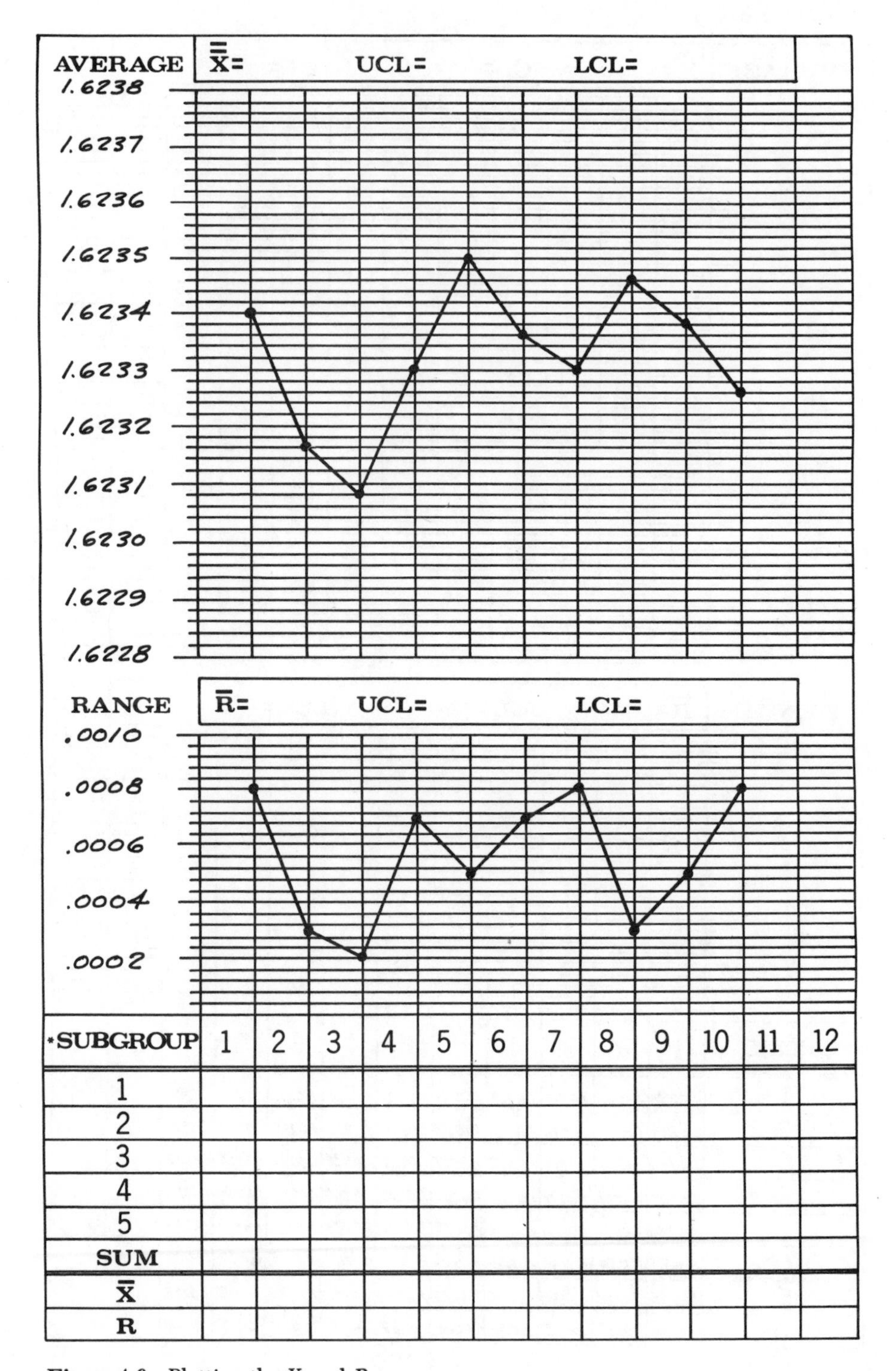

Figure 4-6 Plotting the X and R

FOR THE AVERAGE CHART

$$UCL = \overline{\overline{X}} + (A_2 \times \overline{R})$$
$$LCL = \overline{X} + (A_2 \times \overline{R})$$

FOR THE RANGE CHART

$$UCL = D_4 \times \overline{R}$$
$$LCL = \text{No LCL for subgroup size of 5}$$

Where

UCL = Upper Control Limit	$\overline{R}$ = Average of Subgroup Ranges
LCL = Lower Control Limit	A_2 = Control-Chart Factor (0.58, n = 5)
$\overline{\overline{X}}$ = Grand Average	
$\overline{X}$ = Subgroup Average	D_4 = Control-Chart Factor (2.11, n = 5)

Calculated upper and lower control limits can then be placed on the average-and-range charts in dotted-line fashion, as shown in Figure 4-8.

If all subgroup plot points fall randomly between the control limits on both the average chart and the range chart—that is, if the points have no pattern—the process is considered to be under a state of statistical control. However, if just one plot point falls outside of either chart's control limits, the process is out of statistical control, and one or more variables (problems) must be corrected. Changes or corrections of any sort dictate the collection of additional sampling data.

CAPABILITY VERSUS TOLERANCE

Having once established statistical control on a given process, the engineer can now see if the process is capable of holding the required tolerance. The capability of a process is usually defined as six standard deviations (6S) of the subgroup variation (see Fig. 4-9). The standard deviation can be estimated by using the $\overline{R}$ value from the control chart and the formula shown below:

$$S = \frac{\overline{R}}{d_2}$$

Where

S = Estimated Standard Deviation
$\overline{R}$ = The Average of the Subgroup Range
d_2 = Constant of 2.326 for a Subgroup of 5

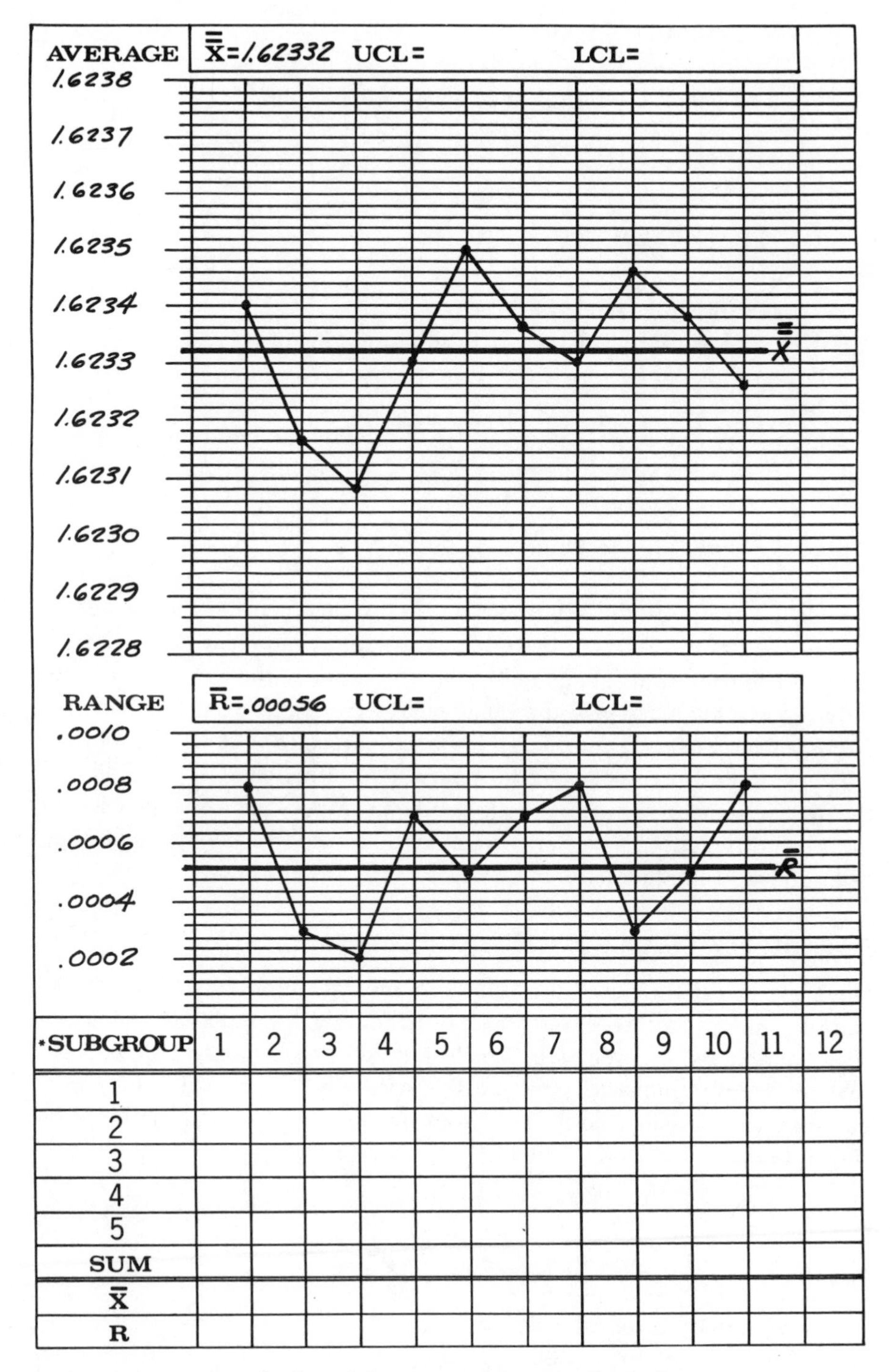

Figure 4-7 Entering the Grand-Average and Average-Range Lines

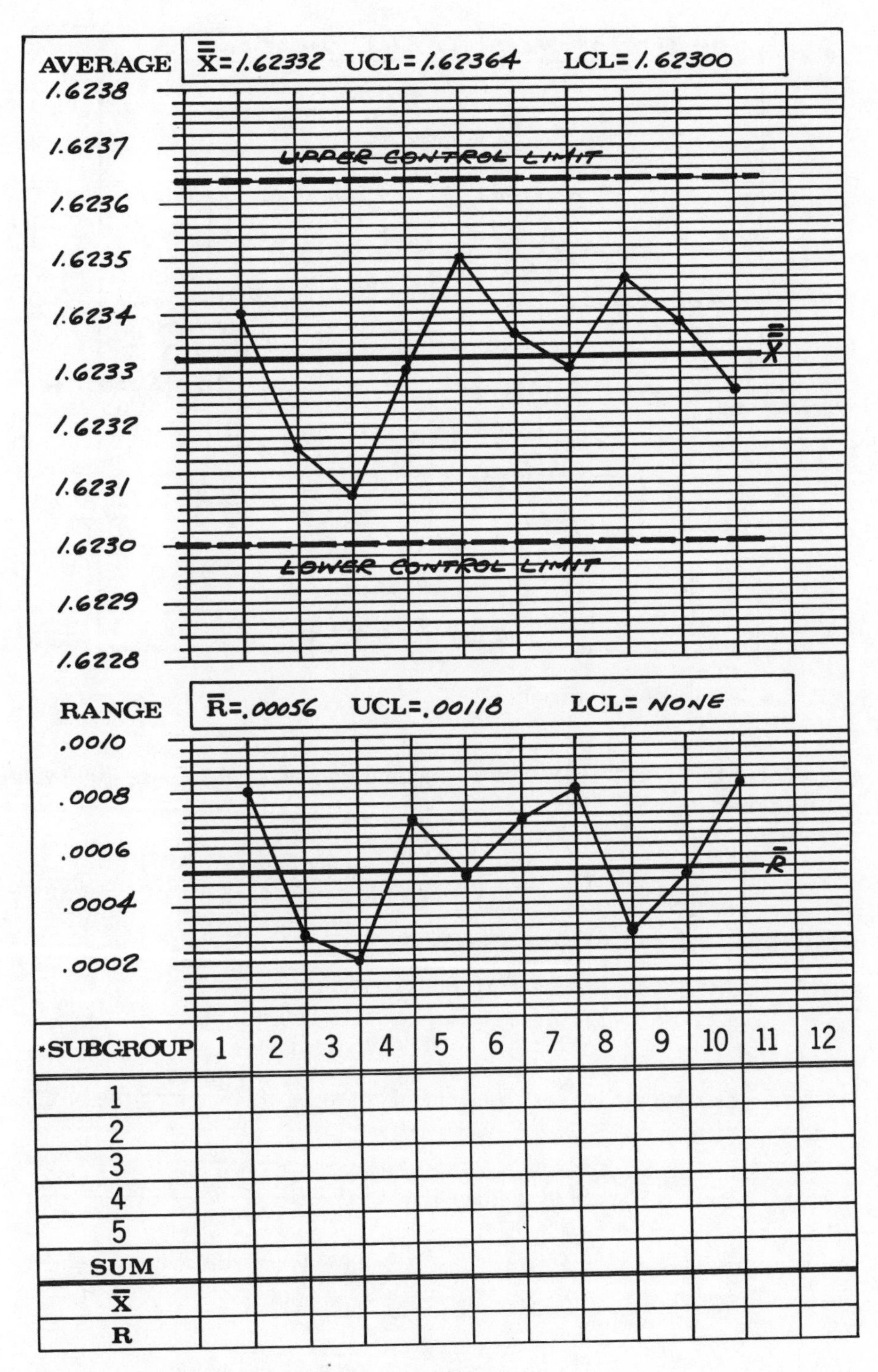

Figure 4-8 Upper-and-Lower Control Limits

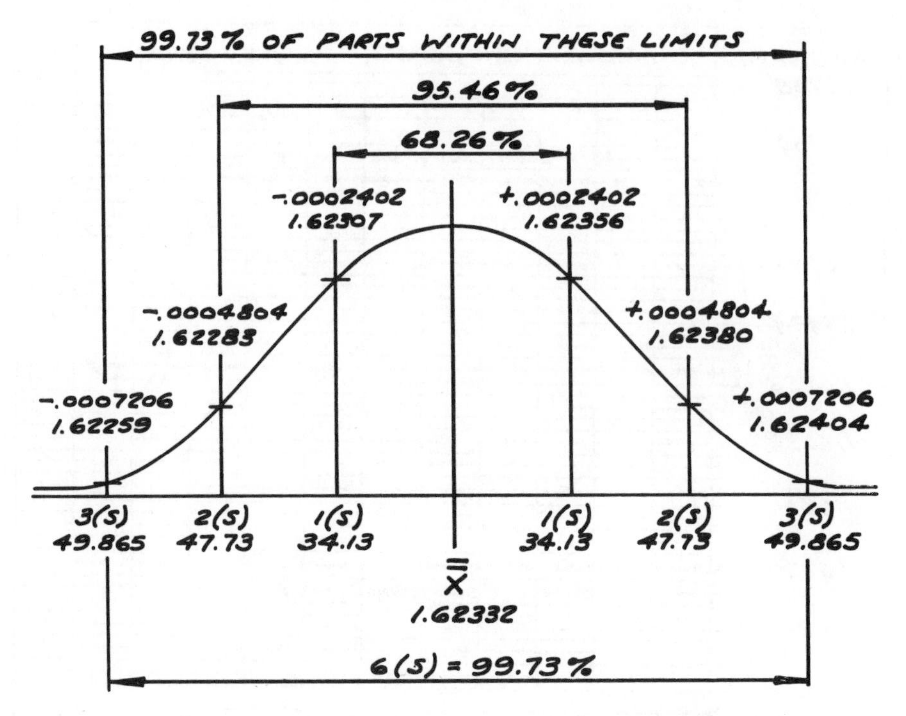

Figure 4-9 Three Sigma Limits

Using the 1.6233 dia. ± .0010 bearing-pocket example, the following standard deviation is calculated:

$$S = \frac{\bar{R}}{d_2} = \frac{.00056}{2.326} = .0002407$$

The process capability is then calculated as:

$$\begin{aligned}\text{Process Capability} &= 6S \\ &= 6 \times .0002407 \\ &= .0014442.\end{aligned}$$

The process capability, as a percent of the print tolerance, is then calculated:

$$\begin{aligned}\text{Process-Capability Percentage} &= \frac{\text{Process Capability} \times 100}{\text{Tolerance}} \\ &= \frac{.0014442 \times 100}{.002} \\ &= 72.21\%\end{aligned}$$

A process capability of 75% or better is generally considered acceptable. It actually means that 99.73% of the time, all parts run will fall within a range only three quarters the size of the print tolerance. Some companies strive to continually lower this process-capability percentage so that less inspection and machine adjustment will be required.

4 ▪ 3 TOOLING COSTS

After the machinery for each operation has been selected, the process engineer must turn her or his attention to that of tooling. Here, the term *tooling* is used to describe all additional items required in the beginning of any given operation, including all durable tools, one set of perishable tools, gauges, and other peripheral items such as tables, racks, bins, conveyors, and so forth.

Tooling and its cost are first considered when the original cost estimate is made. It is recorded on the cost-estimate transmittal (see Fig. 4-1, previously shown). Note how all tooling is lumped together on one line, under a single heading, to save room. A detailed list of the tooling required for each operation should be kept with the CET back-up data (see Fig. 4-10).

The key is not to forget anything, because the actual costs must be kept within the original estimate when and if the contract is won. Therefore, the accuracy of this estimate is extremely important. When treated scientifically, accurate estimates of cost will be yielded. However, rarely will the process engineer be afforded the time to employ sophisticated estimating techniques. A great deal of experience is generally required to make accurate estimates of cost in the time allotted. This section is devoted to outlining a few quick estimating methods that can be utilized by the inexperienced engineer.

It is generally conceded that estimates within plus or minus 10% of the actual costs incurred are acceptable. Management's decision to set prices and offer a quote on work will in large part be based upon this estimate and its assumed level of accuracy. In an effort to improve the quality of original estimates and the work orders they generate, most companies now monitor and publish estimated versus actual costs by work order number and author. The circulation of this record helps to bring real world tooling costs into perspective for the manufacturing engineer and tool designer. Of course this method of cost estimate improvement is inherently tainted, as future estimates will reflect past charges, while never knowing the validity of those charges. To break this chain of inflationary estimating, accurate estimates must be made and enforced.

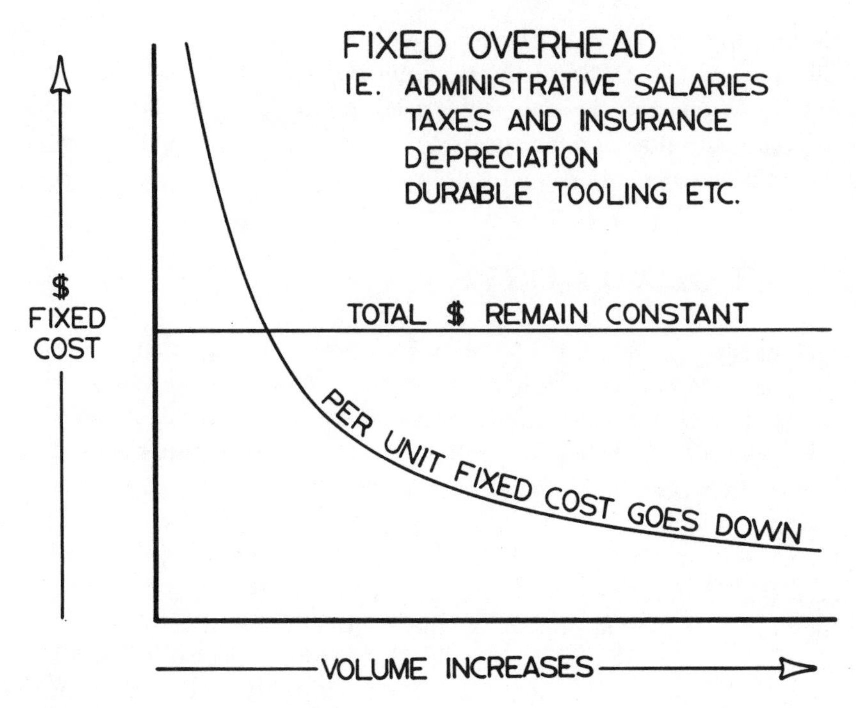

Figure 4-10 Fixed Overhead

Estimating Design Time and Cost

Tool designers are normally required to account for their time by charging labor to one or more work orders in their possession. From the time the designer picks up the work order until the time the design is approved, hours and therefore dollars are being charged against the work order.

Base wages for tool designers typically run from $5.00 to $15.00 an hour depending upon the area of the country. In addition to base wages, fringe benefits and departmental overhead of approximately 50% must be factored in. Total design charges will then run from $7.50 to $22.50 an hour. Checking time must also be considered and will normally equal 25% of the design time. These figures, it should be noted, are for reference use only. Actual company policy should be checked when on the job.

With the hourly cost firmly in mind, the engineer is confronted with the tougher job of estimating how long it takes to design a given piece of tooling. Many rules of thumb exist to aid in this decision-making process. One such rule is based on drawing size. A "D" size (24 × 36 in.) sheet of vellum takes 8 hours to fill, a "C" size (18 × 24 in.) sheet takes 4 hours, and a "B" size (9 × 18 in.) sheet takes 2 hours. Another rule claims one hour of drafting and design time should be allotted for each detail listed on the assembly drawing. These rules can be used with varying degrees of success, provided that the complexity of the design is taken into account. Other factors affecting design time might be the requested tooling's similarity to past designs and the utilization of computer-aided design equipment.

The chief tool designer can be a fine source of information about the design time required for various types of tooling. The manufacturing engineer should retrieve twenty or more accounts of actual design times for future reference. These actual accounts of design time can then be used as benchmarks while additional estimating experience is gained.

Estimating Build Time and Cost

Tool and die makers must also account for their time by charging labor to one or more work orders. Because of the machinery used in tool and die making, departmental overhead costs run much higher than those in the design department. Total toolmaking charges will typically run from $15.00 to $30.00 an hour. The tool and gauge inspection labor required to check the tooling will also be charged to the work order.

Estimating the hours required to build a given tool is extremely difficult. This difficulty stems from two sources: (1) the design has not been completed at the time that the estimate is being made and (2) the process engineer does not have the time, in most cases, to make in-depth calculations about metal removal rates and in-house machining capabilities. A standard data library of actual tool build times should be created to enhance the accuracy of future estimates.

Estimating Material Cost

Figuring material costs makes up the final component in the work-order estimating process. When necessary, available material sizes and exact costs can be acquired by phone through local suppliers. If reasonably close estimates are acceptable, price lists given in recent mill supply catalogs will provide sufficiently accurate information. The same rationale is acceptable and can be used for estimating the cost of standard, commercially available, tooling components.

The Volume Effect and Cost

The estimate of cost presupposes that the tooling being requested is economically justifiable. This may or may not be the case. The degree of tooling sophistication that can be economically produced depends upon one or more of the following factors:

1. The required production rate per hour (The pieces produced in a given period of time.)
2. The product stability (What length of time will the product be in production?)
3. The available machinery (Is it new or used, shared, used for more than one part, or dedicated, used for a single part?)
4. Direct costs (The hourly wages plus fringe benefits.)
5. Indirect costs (The setup time, start-up, and shake-down, utilities, and so forth.)
6. Part tolerances (Dimensional limits and surface finishes.)

These factors should have been previously considered by the manufacturing engineer, but may also be questioned by the tool designer. (Curtis, *Tool Design,* 1986)

4 ▪ 4 PRODUCTION RATES

Hourly rates of production must also be established for each operation during the cost-estimating stage (again, see Fig. 4-1, previously shown). These rates can be derived through longhand methods, historical data, micromotion study, or by utilizing a microcomputer-based system such as the one produced by MICAPP Incorporated and marketed through The Society of Manufacturing Engineers.

Again, an accurate estimate of potential hourly production is imperative. It impacts the sale price of the product, the number of machines necessary to meet required production volumes, machine utilization figures, and production scheduling. As with other types of estimates, an accuracy of ± 10% is considered acceptable. These estimated production rates will stay in force until industrial engineering can time-study actual setups and establish a truly accurate production rate for each operation.

Direct-Labor Content

Direct labor is the actual cost of labor required to process parts. Any labor that changes the part or otherwise adds to its value is considered direct labor. For example, cutting bar stock into slugs for further processing is a direct-labor operation, but transporting the slugs to the next operation is not. In this case, the transportation is considered indirect labor and will fall under the category of overhead.

The per-part, direct-labor content for any one operation is figured using the following formula:

$$\text{Per-Part Direct Labor} = \frac{\text{Hourly Rate of Pay}}{\text{Hourly Rate of Production}}$$

EXAMPLE

$$\text{Per-Part Direct Labor (For Bar-Stock Cut-Off Operation)} = \frac{\$13.85 \text{ (Hourly Rate of Pay)}}{200 \text{ (Hourly Rate of Production)}}$$

$$= \$0.06925$$

Therefore, in this example, a direct cost of almost seven cents per part can be attributed to the cut-off operation. Other direct-labor operations required in the manufacture of this or any part would be similarly calculated and added together, yielding the part's *total direct-labor content*. These direct-labor dollar values are then added to the CET (see Fig. 4-2, previously shown).

The direct labor is only a portion of the total cost required to produce a given part. However, all other costs, such as overhead, are expressed as a percentage of the direct-labor content.

For example, if a part's direct-labor content was $1.00, the overhead might be expressed as 300%, 3 × $1.00, or $3.00. This means the total cost to produce the part would be $4.00: direct labor ($1.00) plus overhead ($3.00).

Overhead rates in industry have commonly run between 200% and 1500% of direct labor, with highly automated operations having high rates of overhead compared to their direct-labor costs.

Total overhead, also referred to as *burden*, is further broken down by accounting methods into fixed and variable categories. *Fixed overhead,*

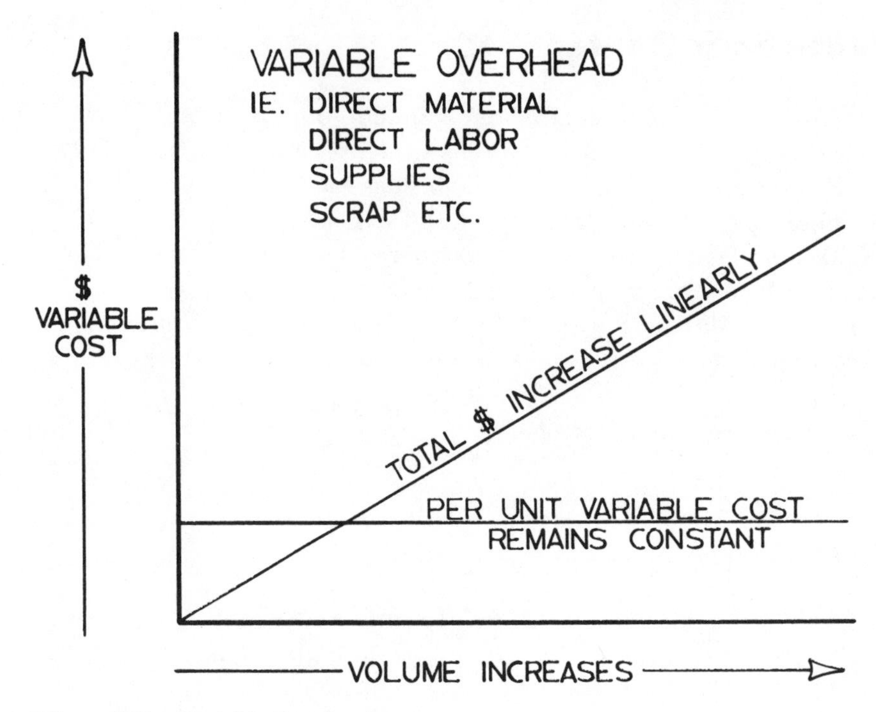

Figure 4-11 Variable Overhead

such as administrative salaries, property taxes, and so forth, is made up of those items that tend to remain independent from increases or decreases in total production volume (see Fig. 4-11). On the other hand, *variable overhead,* which includes indirect labor, perishable tools, direct material, and the like, is made up of those items that tend to identify or vary with total production volumes.

4 ▪ 5 CONCLUSION

Chapters 2, 3, and 4 served to walk the engineer through those concerns that must be addressed in the cost-estimating stage of process planning. At this point, the planner can forward and file the cost-estimate transmittal (CET) with confidence. If a contract is won, the CET will now provide an excellent point of departure in moving toward the implementation of the formal process plan.

REVIEW QUESTIONS

1. What are some of the major economic processing considerations that must be addressed by the process planner?
2. What are the factors affecting proper machine selection?
3. What is a break-even chart used for?
4. What is a payback comparison?
5. List some benefits that may be realized through the use of existing machinery.
6. List some problems in selecting existing machinery for use in new process plans.
7. What is meant by the term "machine capability"?
8. Why is an accurate estimate of each operation's production rate imperative?
9. What is direct labor?
10. What is fixed overhead? What is variable overhead?

CHAPTER 5

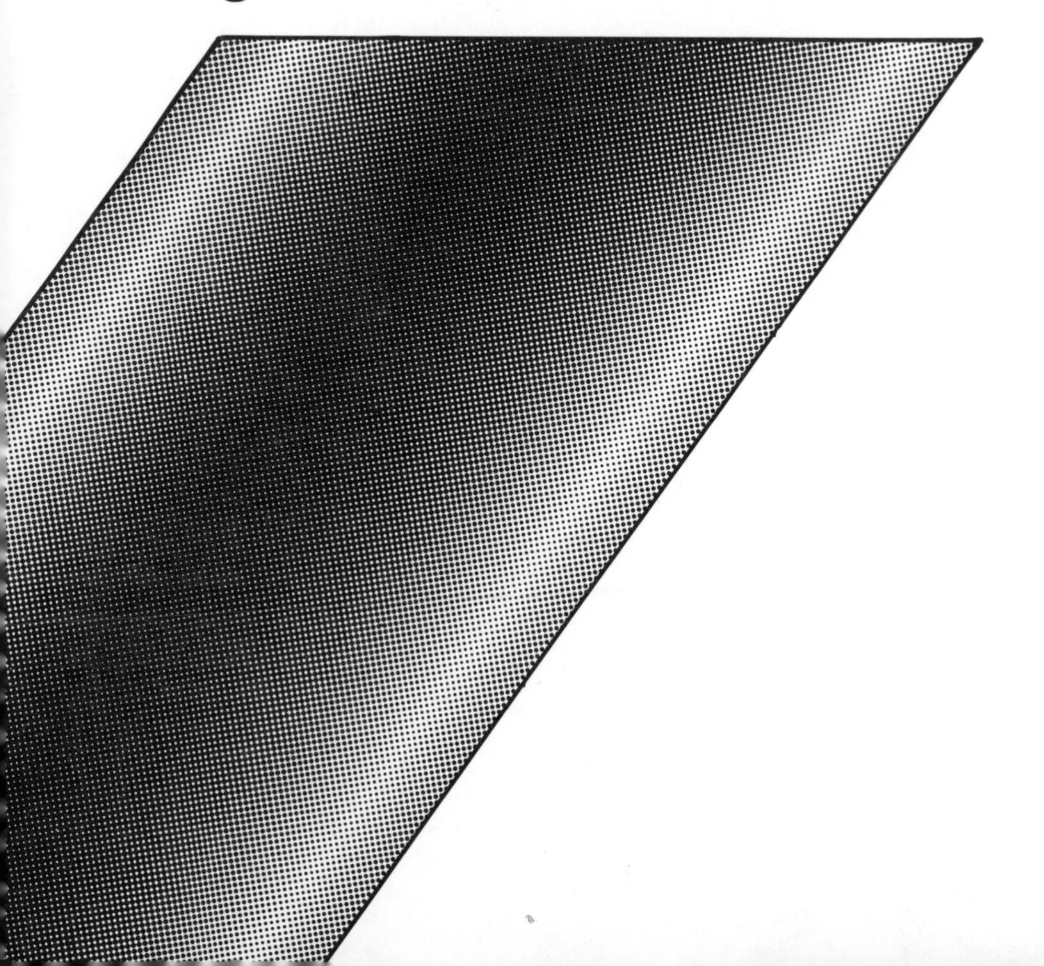

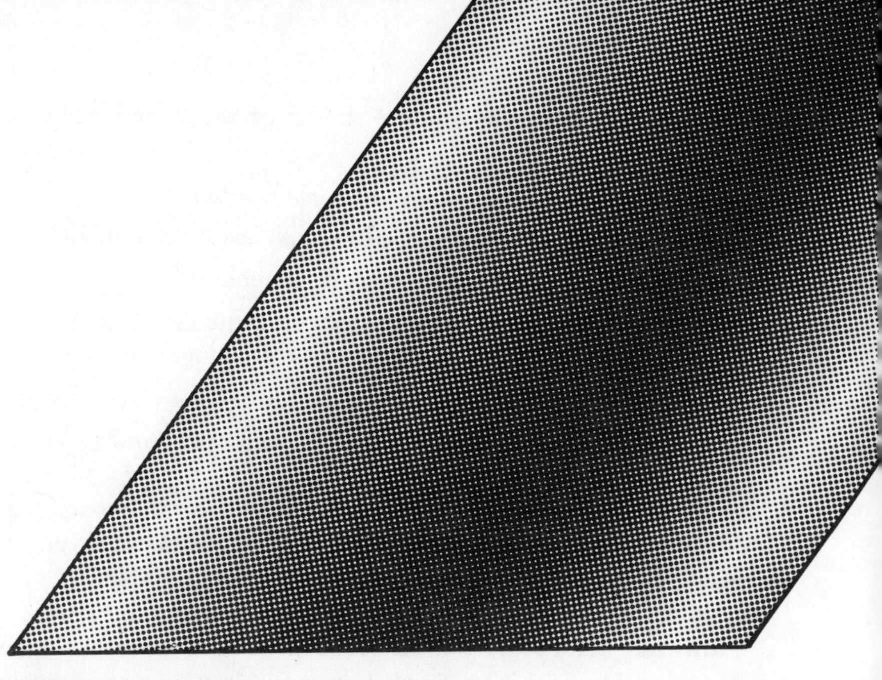

Process Documentation

5 • 1 INTRODUCTION

Manufacturing may be considered the production of any item in accordance with an organized plan. The organization and detailing of this production plan is the charge of the process engineer.

As the process engineer looks at a new product, many production decisions must be made. This decision-making process is referred to as process planning. Some of the more important process-planning questions are as follows:

1. Which details in the product should be made and which should be bought?
2. What raw material is required for each detail?
3. Which operations are required to make each detail of the product?
4. How should selected operations be sequenced?
5. What dimensions and tolerances are required at each operation to assure conformance with the product drawings?
6. What kind of capital equipment is required for each operation?
7. What type of tooling is required for each operation?

Once these process-planning questions and others have been sufficiently answered, they must be documented in the form of processing drawings. Process drawings then become the single most important source of information in the manufacturing plant. (Curtis, *Tool Design*, 1986)

However, before any time is spent on process documentation, formal approval to do so must be granted through some type of internal releasing authority.

5 ▪ 2 RELEASE AUTHORITY

In most manufacturing companies, cost estimates are provided for hundreds of potential jobs, which are bid upon each year. Only a small percentage of those bids will end up yielding formal production contracts. Therefore, to prevent a waste of engineering time and tooling money on jobs (contracts) not yet won, management typically uses a *formal release document* (see Fig. 5-1) to authorize the expenditure of company resources on a given job.

Without this formal release authority, the process engineer usually proceeds only as far as the cost estimate. In an effort to get a head start on a large processing and tooling project, the engineer may sometimes be instructed to begin formal processing work without the benefit of a formal release. This is of course a gamble of time and money, which the company loses if a production contract is not forthcoming.

5 ▪ 3 COST-ESTIMATE TRANSMITTAL

Upon receiving formal release authority, the process engineer pulls and reviews the original cost estimate with all its related information (see

RELEASE AUTHORIZATION

☐ RELEASE FOR PROCESS PLANNING
☐ RELEASE FOR TOOLING & GAGES
☐ RELEASE FOR PRODUCTION

AUTHORIZATION
A. SALES________________
B. MARKETING________________
C. GEN. MGT.________________
D. CONTROLLER________________

RELEASE INFORMATION:

1. PRODUCTION SAMPLES REQUIRED:
   DATE____________
   QUANTITY__________
2. PRODUCTION PARTS REQUIRED:
   DATE____________
   QUANTITY/YEAR______________
3. AUTHORIZED EXPENDITURES:
   CAPITAL EQUIPMENT____________
   DURABLE TOOLING______________
4. LENGTH OF PRODUCTION CONTRACT_________
5. PART NUMBER:____________________
6. CET NUMBER:_____________________

Figure 5-1 Internal Releasing Authority

Fig. 5-2). This review must be made in light of any changes that have taken place between cost estimate and production release. Such changes might include dimensions, volume requirements, and production deadlines. In spite of these changes, the process planner must often live within the tooling dollars originally quoted.

Changes to the product itself will usually be documented through the use of an engineering change notice.

The Engineering Change Notice

An Engineering Change Notice (ECN) is a formal document that authorizes a change to be made on an existing product drawing (see Fig. 5-3). It states, in both word and drawing form, exactly what is to be changed and for what reasons. Although the ECN is not actually a product drawing, it affects product drawings to such an extent that a review here is necessary.

Once any product design is released to sales, purchasing, and process engineering, it is considered to be in final form. This final form simply means that one workable design solution has been agreed upon. For many

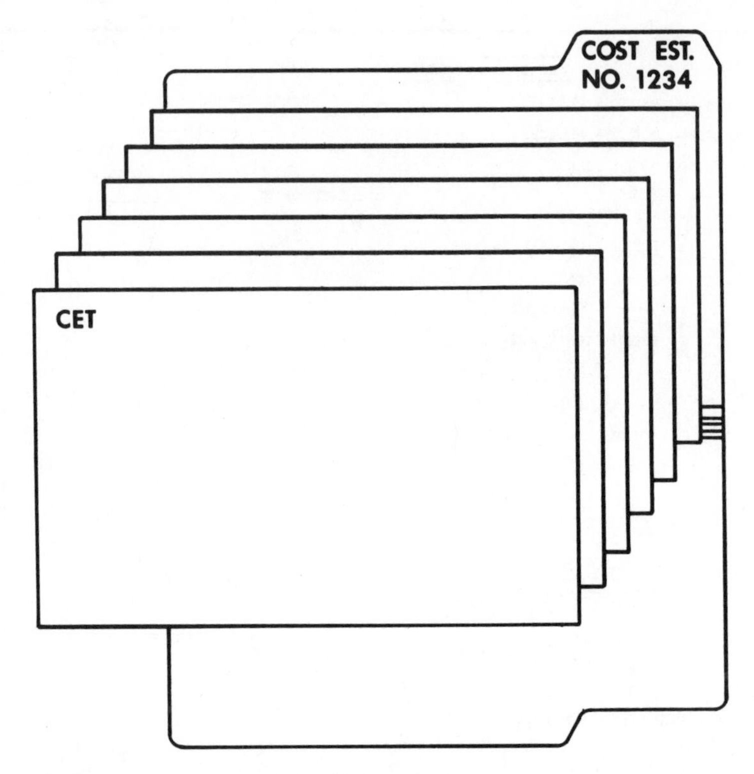

Figure 5-2 Cost Estimate Revisited

reasons, a design change may be required after the design is released to production. Some of the more common reasons for these design changes are listed below:

1. To correct an error
2. To facilitate manufacture
3. To improve product performance
4. To minimize product cost
5. Customer request
6. Standardization
7. Safety or product liability

Many times a single design change will have far-reaching implications in terms of the time and cost required for implementation. When this is the case, a document referred to as an Engineering Change Request

(ECR) precedes the generation of an ECN. The ECR must be approved by process engineering and other affected departments before an ECN becomes formal.

Much of the manufacturing, engineering, and tool-design work completed for existing products is originally generated by the ECN. The importance of the ECN cannot be overemphasized, and the process-engineering and tool design departments must give high priority to the completion of all work dictated by the ECN. (Curtis, *Tool Design,* 1986)

5 ▪ 4 THE ROUTING

Most details, subassemblies, and final assemblies travel through a series of individual manufacturing operations on the way to completion. The routing is a prearranged list of the order in which these operations are to be executed.

ECR _______ ENGINEERING CHANGE ECN _______

x	Reason For Change	DETAILS	Date
	Decrease Cost		Origin
	Improve Quality		Part No
	Correction		Name
	Customer Request		Customer
	Mandatory		Dwg Chg Ltr
	Facilitate Mfg		Supersedes
	Other		Service Aff
			Stock Disposition
			Approved By

Process Remarks	Cost	Pres	Prop	Diff
	Material			
	Labor			
	Burden			
	Total			
	Tools			
	Equip			

Production Remarks	
	With Obsolescence
	Without

Sales Remarks	ECN	COPY	INT
		Prod Mgr	
		Mfg Eng	
		Sales Mgr	
		Sales Eng	
		Superintendent	
		Eng Mgr	
Effective	Issued By	Product Eng	
Date Issued	Approved By	Originator	

Figure 5-3 The Engineering Change Notice

The routing is normally sheet one of the process drawings and provides the following information (see Fig. 5-4):

1. Part name and number
2. The department number in which each operation takes place
3. A brief written description of each operation
4. The machine class of each operation (Machine class relates to the operator's rate of hourly pay.)
5. Process sheet numbers
6. Operation numbers
7. The hourly production rate or standard
8. The base rate of pay (This does not include fringe benefit costs.)
9. Accumulated direct labor costs
10. The setup hours required for each operation
11. A record of any changes

Note how the operation numbers are assigned in multiples of ten (i.e., 10, 20, 30, etc.) and how lines have been left blank on the routing between each operation. This operation recording method allows for the addition of new or forgotten operations without disturbing the numbering of previous operations.

NAME GEAR DATE 8-10-83 PART NO 40334
ASSY NO 15531 CUST G.M.
MATL SAE-4140 SHEET NO 1

Dept	Description	Sheet	Oper	Hourly Production	Rate	Labor	Set-Up Hours
102	MACHINE GEAR SIDE	2	10	125	9.45	.076	4
102	MACHINE HUB SIDE	3	20	125	9.45	.076	4
102	BROACH INTERNAL SPLINE	4	30	150	9.20	.0613	2
102	CUT TEETH	5	40	75	9.45	.126	4
102	SHAPE SPLINE	6	50	50	9.45	.189	1
113	HEAT TREAT	7	60	1000	10.75	.0107	—
102	GRIND HUB	8	70	125	9.85	.0788	2

Date	Change	By	Date	Change	By	Date	Change	By
9-5-83	RELEASED TO PRODUCTION	M.C.						

Figure 5-4 Routing

Base operation numbers are assigned to operators, not to machines. For example, at an operation number 10, one operator might run three different machines simultaneously: a lathe, a milling machine, and a drill press. The assignment of operation numbers, in this case, would be as follows: the lathe work is operation 10A, the milling machine work is operation 10B, and the drilling is operation 10C. This numbering scheme is dictated by material handling and production control procedures. (Curtis, *Tool Design*, 1986)

The Tolerance Chart

Before the optimum sequence of operations can be determined and placed on the routing, the process engineer must make a tolerance chart (see Fig. 5-5). In brief, the tolerance chart provides the engineer with a systematic way to determine and maximize dimensional tolerances at each operation; this is accomplished through proper sequencing of all operations.

Due to the complexity of this subject, the entire content of Chapter 6 has been devoted to the explanation of tolerance-charting techniques.

5 ▪ 5 OPERATION SHEETS

An operation sheet, also called a process sheet, is a complete set of instructions utilized by setup, production, and inspection personnel in the normal course of their work.

KEY CONCEPT

Each operation listed on the routing will require an operation sheet.

Operation Sheet Information

The following information should be provided on each operation sheet (see Fig. 5-6):

1. General title block information (for example, part name and number, operation description and number, date, and so on)
2. An orthographic drawing of the part as it appears after the operation

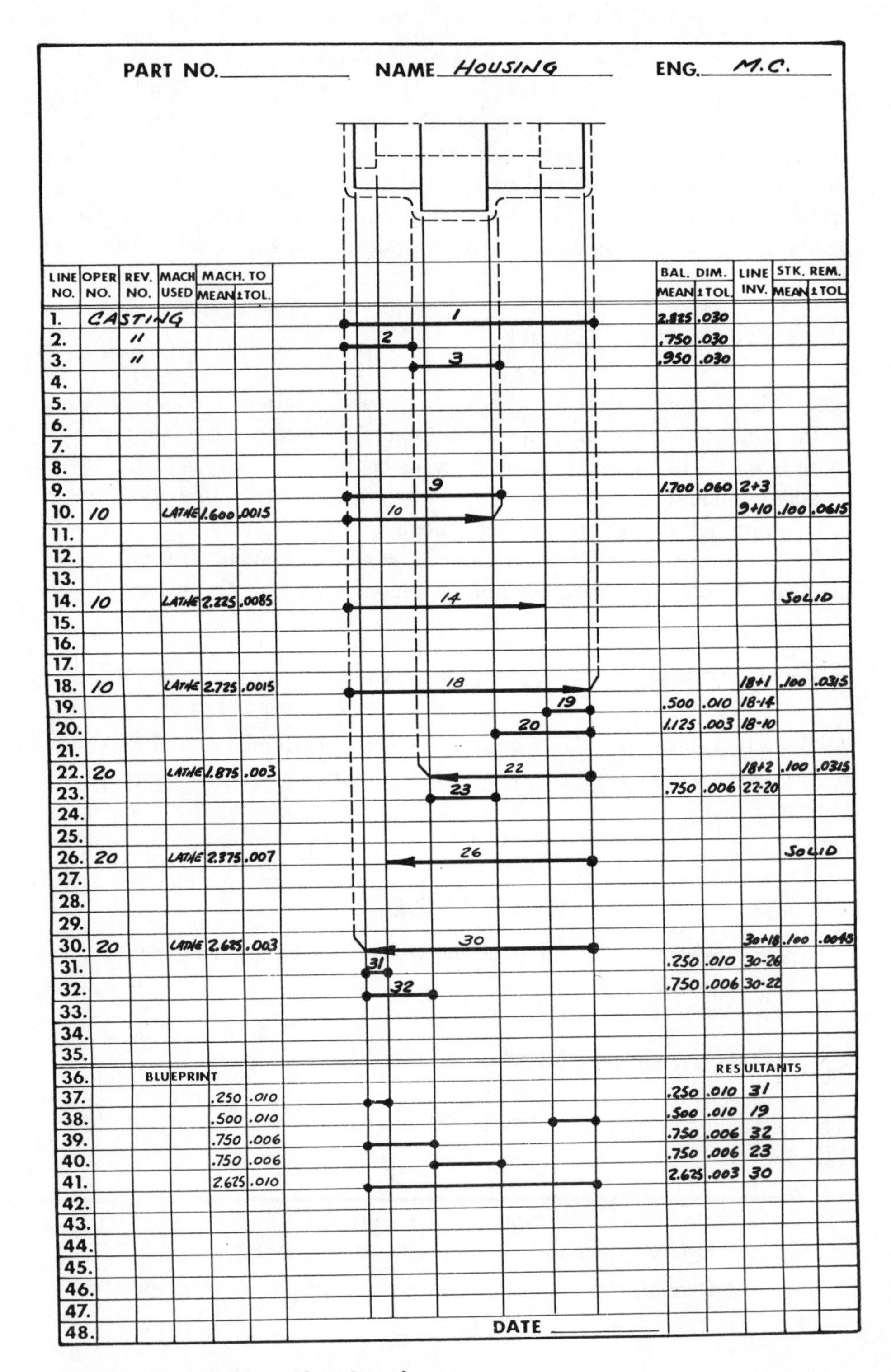

Figure 5-5 Tolerance Chart Sample

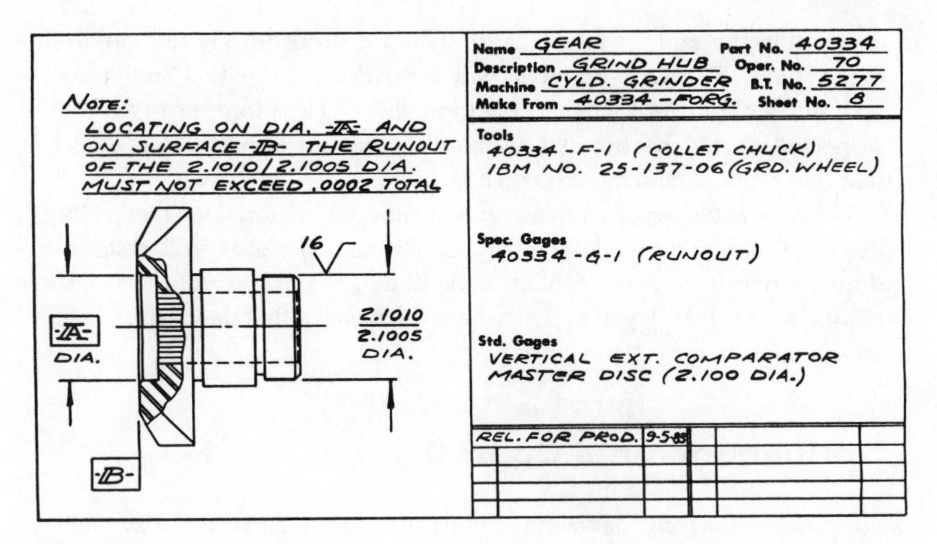

Name GEAR Part No. 40334
Description GRIND HUB Oper. No. 70
Machine CYLD. GRINDER B.T. No. 5277
Make From 40334 - FORG. Sheet No. 8

NOTE:
LOCATING ON DIA. -A- AND
ON SURFACE -B- THE RUNOUT
OF THE 2.1010/2.1005 DIA.
MUST NOT EXCEED .0002 TOTAL

16
-A-
DIA.
2.1010
2.1005
DIA.
-B-

Tools
40334-F-1 (COLLET CHUCK)
IBM NO. 25-137-06 (GRD. WHEEL)

Spec. Gages
40334-G-1 (RUNOUT)

Std. Gages
VERTICAL EXT. COMPARATOR
MASTER DISC (2.100 DIA.)

REL. FOR PROD. 9-5-85

Figure 5-6 Operation Sheet

3. Dimensions and tolerances relating to the operations
4. Locating- and clamping-points
5. A list of all special and standard tooling required at the operation
6. A list of all special and standard gauges required, including inspection frequencies
7. Release information
8. Change block information

There are two distinct levels of the operation-sheet release procedure. The first, "Release for Tools and Gauges," means the process has been firmed up to the point where the design of tooling can begin. Under this type of release, the tool-design department is the only group in the plant authorized to use the information contained on the operation sheets. The second level, "Release for Production," means that the information contained on the operation sheets is now authorized for use on the production floor.

Tool-Design Feedback

Operation sheets are drawn up by process engineers as part of their total process-planning routine. In most cases, these operation sheets are not checked in the same manner that is standard procedure in the tool-design department. The tool designer will normally be the first individual to take an in-depth look at the operation sheet. Therefore, errors found or suspected should be brought to the attention of the process engineer so that corrective action can be taken.

Other designer/engineer interaction may be required as the tooling is designed and numbered. Many times the tool designer will assign tool-drawing numbers as the tooling is designed. When this is the case, these numbers must be relayed to the process engineer so that they can be entered on the operation sheets.

Nondimensional Operations

Some operations, such as wash, deburr, heat treat, and paint, have no specific dimensions. These nondimensional processes will still require individual operation sheets to outline procedures and equipment.

5 ▪ 6 TOOL DRAWINGS

The term *tooling* is used in reference to all hardware employed in the manufacture of consumer goods. Tool drawings, in various forms, are semipermanent records of this hardware.

These are working drawings of durable and/or perishable tooling that require machining, forming, welding, special assembly, or any other method of alteration. The alteration of standard tooling components or the design of special tooling are the only reasons for the generation of tool drawings.

Assembly and detail drawings of the same size and tool number make up the typical tool drawing (see Figs. 5-7 and 5-8).

5 ▪ 7 TOOL LAYOUTS

A tool layout is a drawing that shows the arrangement of all tooling used in a specific operation or setup.

One type of tool layout, sometimes referred to as a machine layout, is developed as complex tooling systems are designed for automatic machinery. This tool layout is used to show the relationship of machine components and tooling to the space available. All tooling in the setup is shown in the

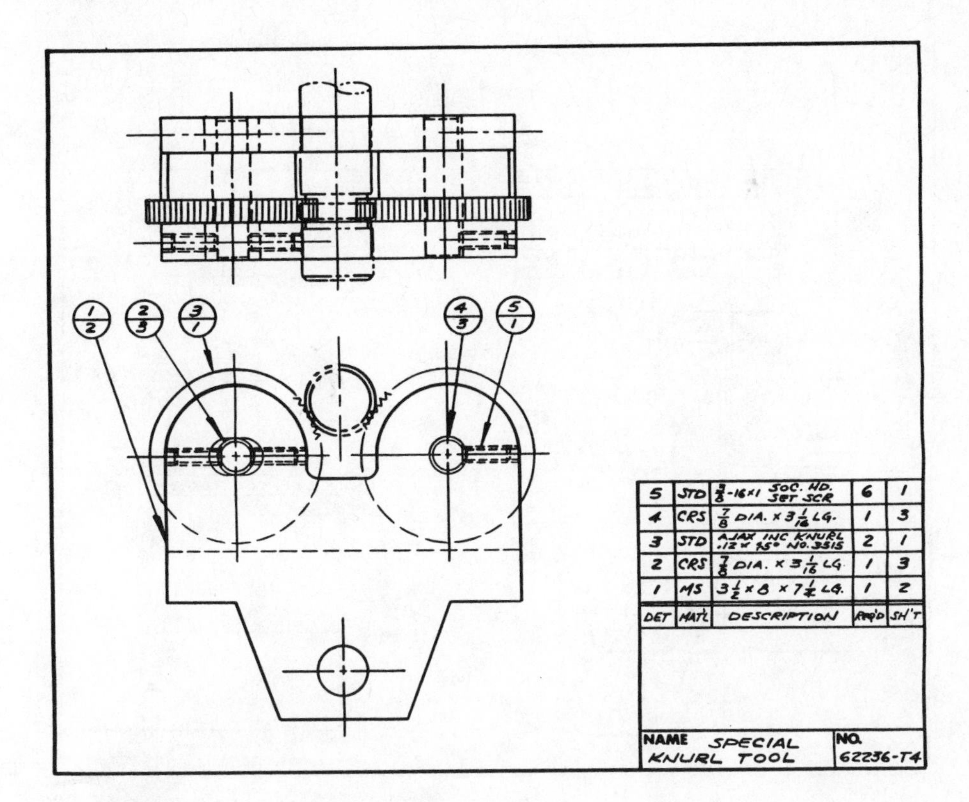

Figure 5-7 Tool-Drawing Assembly

extreme positions of movement (see Fig. 5-9). This illustration of tooling interaction is used by the tool designer for reference purposes. Individual tools are designed in accordance with the clearances dictated by the tool layout. This style of tool layout is often drawn in full scale to ensure a proper representation of the hardware involved. The sheer size of the layout normally will make it too cumbersome for use out on the shop floor.

A second type of tool layout, commonly called a strip or setup chart, is drawn after all the tooling required in the operation has been designed. The strip chart gives a pictorial illustration of all tooling required in each station of a machine as it performs work on a given part (Fig. 5-10). In addition to the pictures, tool numbers, speeds and feeds data, and the sequence of the operation are also listed. To make this setup information suitable for use on the shop floor, it is normally confined to one sheet of paper. To accommodate the placement of this information on one sheet, required drawings usually are not made to scale.

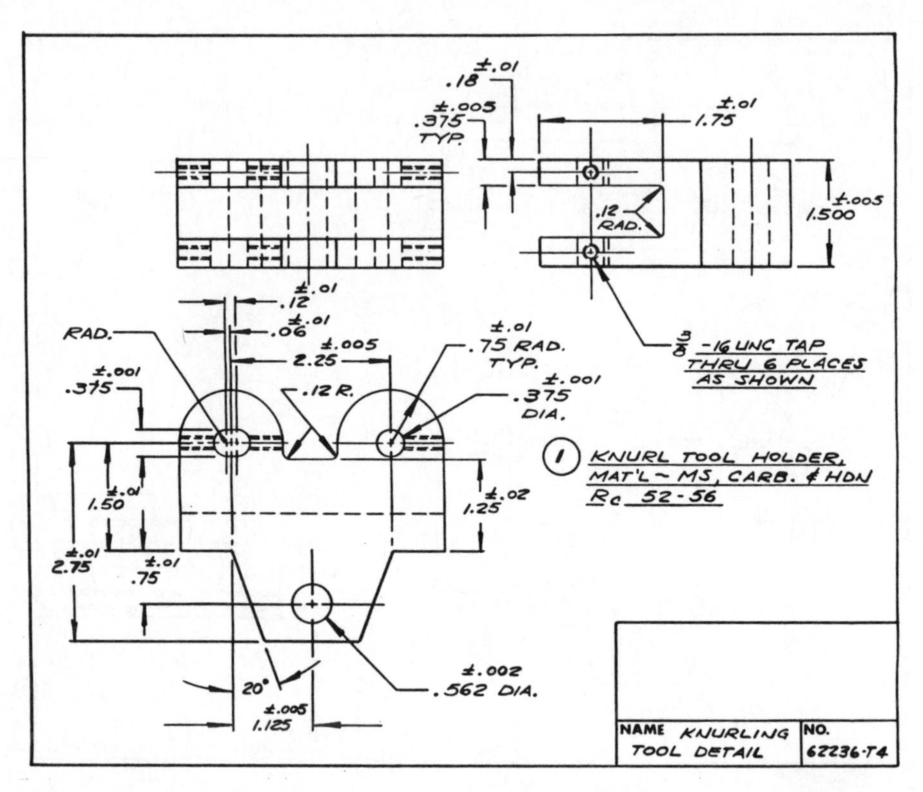

Figure 5-8 Tool-Drawing Detail

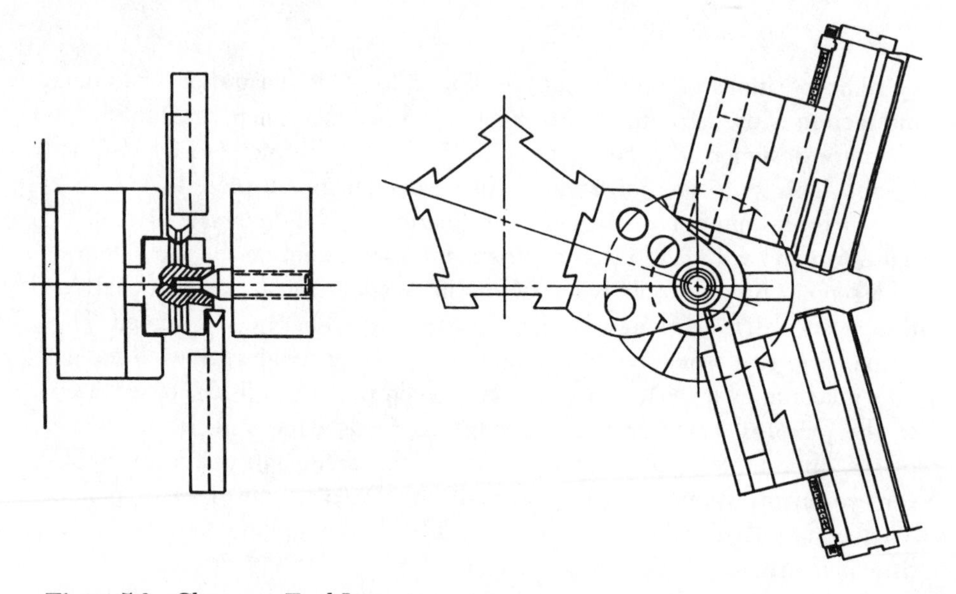

Figure 5-9 Clearance Tool Layout

NAME BEARING RACE DRAWING NO. 93471-L1

DESCRIPTION ROUGH MACHINE B.T. NO. 1547

MACHINE 2 5/8 RB-6 ACME DATE 1-3-87

Position / Operation	Qty.	Tooling
6TH POS. STOCK ADVANCE	6 6 6 1	COLLETS PUSHERS SPOOL BUSHINGS STOCK STOP
1ST POS. FACE END	1 1	FLAT FORM TOOLHOLDER STD. CARBIDE TL.
2ND POS. SPOT DRILL FORM O.D.	1 1 1 1 1	DOVETAIL FORM TOOLHOLDER FORM TOOL DRILL DRILL CHUCK COLLET
3RD POS. DRILL TO DEPTH	1 1 1	DRILL DRILL CHUCK COLLET
4TH POS. CHAMFER ID. START CUT-OFF	1 1 1 1	CUTOFF TOOL CUTOFF TLHLD. RECESS TOOL RECESS FIXT.
5TH POS. CUTOFF	1 1	CUTOFF TOOL CUTOFF TLHLD.

Figure 5-10 The Set-up Chart

KEY CONCEPT

Each operation sheet should have a corresponding tool layout.

5 ▪ 8 REFERENCE DRAWINGS

A reference drawing is any drawing or sketch used by the tool designer or process engineer to solve a tooling problem. These reference drawings will be filed along with the checker prints and will serve in the future as a testimonial to the decisions made at the time of the design release.

Exploded Assembly Drawings

The exploded assembly drawing is an isometric illustration of how details go together to form an assembly (see Fig. 5-11). This pictorial drawing style is much easier for the layperson to understand than orthographic projection.

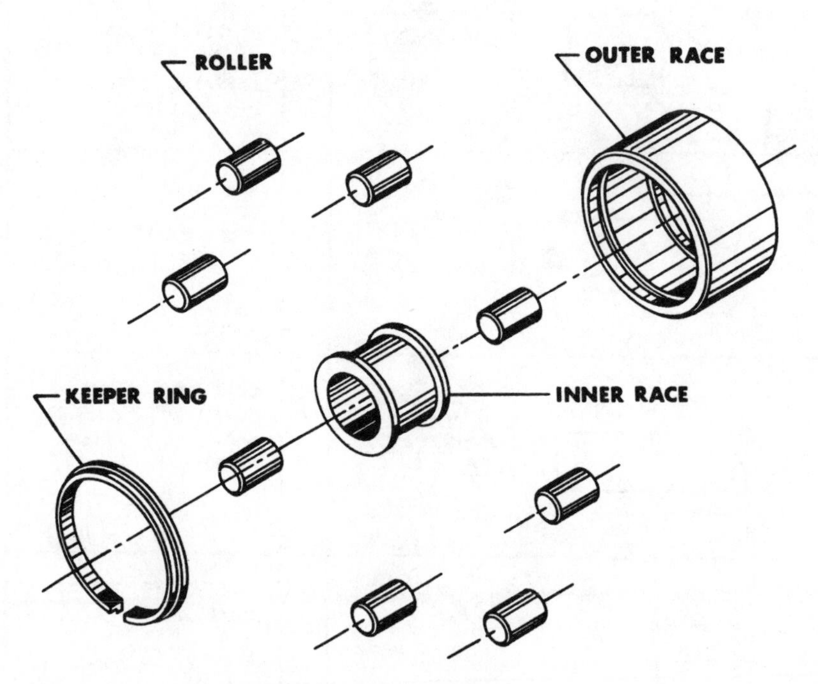

Figure 5-11 Exploded Assembly Drawing

For this reason, exploded assembly drawings have been popularly applied in instruction manuals, service manuals, do-it-yourself kits, and advertising literature.

This drawing style can also be used to help the tool designer visualize complex product asssembly situations. The creation of these drawings is a time-consuming process, normally performed by the product design or technical illustration department. Tooling is rarely if ever drawn in this manner, thus relegating the exploded assembly drawing to reference status.

Sketches

Often made during the preliminary phases of the design process, a sketch is a rough freehand drawing that describes the main features of an object. For years, drafters, designers, and engineers alike have understood the time-saving benefits of the sketch. The completeness and quality of the typical sketch, however, will render it unsuitable for use as a permanent tool drawing. Therefore, sketches must be drawn, considered, and maintained for reference purposes only. (Curtis, *Tool Design*, 1986)

5 ▪ 9 THE DESIGN WORK ORDER

The design work order is a major communication and record-keeping tool used throughout manufacturing. First written by the process engineer, the design work order is a written request for design and/or toolroom labor, including materials and the necessary funds required to cover the same. Once approved, it becomes the authorization for all work outlined on the order to begin. The design work order will typically follow the path listed below:

1. The process engineer writes the order, which includes all necessary information.
2. The appropriate management personnel then approve or reject the work order as written.
3. When approved, the process engineer will receive a copy of the order, while the original moves on to the chief tool designer.
4. The chief tool designer records the receipt of the work order and sets a priority for it to be worked on.

5. When the work order finally reaches the appropriate level of importance it will be assigned to a tool designer.
6. The tool designer will then design the tooling requested, while keeping a running dated record of the design hours spent on the order.
7. The work order, along with the completed design, must then go to the tool-design checker for inspection and the ordering of purchased parts.
8. Once approved, blueprints of the design, along with the work order, will be sent to the toolroom so the tooling can be built.
9. The chief tool designer records the date the design left the tool-design department.
10. The toolroom supervisor will record the receipt of the work order and set a priority for it to be worked on.
11. When a toolmaker is available, the building of the tooling will begin. A running dated record of toolroom labor and material costs will also be maintained.
12. The completed tooling will next be sent to tool and gauge inspection, along with the corresponding blueprints and work order.
13. When approved by tool and gauge inspection, the tooling is staged in a storage area awaiting installation, while the process engineer is notified. If the tooling is not dimensionally correct (to print), the process engineer will be required to intervene.
14. A copy of the work order is sent to the chief tool designer so that a completion date can be recorded. At the same time, the original is sent to accounting for the accumulation of cost.

A minimum of 9 people and 14 separate transactions are involved in the life of the ordinary design work order. Therefore, an intimate knowledge of this document is required of the process engineer and tool designer. This chapter is devoted to the exploration of each piece of information found within the work order. Illustrations of both blank and completed design work orders can be seen in Figs. 5-12 and 5-13, respectively. These figures should be referred to as each section of this chapter is reviewed.

The Tooling Description

The tooling description is a one-line explanatory statement of the tooling to be designed. This abbreviated statement, along with the work order

WORK ORDER			REL. DATE	W.O. NO.
TOOL NAME			REF. TL.	TOOL NO.
OPER. DESCRIPTION			OPER. NO.	DEPT. NO.
PART NAME			PART NO.	ASSY. NO.
MACH NAME			B.T.	COST EST.
NEW DESIGN ☐ BUILD ☐	CHG DESIGN ☐ TOOL ☐	ALSO USED ON:	DATE REQD. DESIGN	SIGNATURE
DATE TO TL. DES.	DATE TO TL. RM.			AUTHORIZED
FROM TL. DES.	DATE TO ACCT.		BUILD	
INSTRUCTIONS:				
COMMENTS:				

Figure 5-12 Blank Work Order

and tool number, is used to reference the design work order in all records, reports, and correspondence concerned with tooling in progress. A few examples of tooling descriptions are listed below:

1. Special 6-in.-dia. milling cutter
2. Brazed carbide boring bar
3. Dovetail form tool
4. Material handling rack
5. Chuck jaws and part stops
6. Profile-grinding dresser cam
7. Paint hook
8. Endurance tester
9. Flush pin gauge
10. Drill jig

WORK ORDER

TOOL NAME	REL. DATE	W.O. NO.
DRILL JIG	12-4-84	14-2378
	REF. TL. 30540-F1	TOOL NO. 32246-F3
OPER. DESCRIPTION: DRILL (12) .250 DIA. HOLES	OPER. NO. 40	DEPT. NO. 28
PART NAME: DIFFERENTIAL HOUSING	PART NO. 32246	ASSY. NO. 15570
MACH. NAME: ALLEN GANG DRILL	B. T. 4227	COST EST. $2300

NEW	CHG.	ALSO USED ON:	DATE REQD.	SIGNATURE
DESIGN ☒ BUILD ☒	DESIGN ☐ TOOL ☐	NONE	DESIGN 2-8-85	[signature]
DATE TO TL. DES.	DATE TO TL. RM.			AUTHORIZED
FROM TL. DES.	DATE TO ACCT.		BUILD 4-15-85	

INSTRUCTIONS: DESIGN AND BUILD (1) PUMP JIG STYLE DRILL JIG TO DRILL (12) .250 DIA. HOLES IN RING GEAR MTG. FLANGE (SEE ATTACHED PRINT FOR DATUMS)

COMMENTS:
THIS IS PHASE I TOOLING

Figure 5-13 Work Order Filled Out

The process engineer formulates the tooling description at the time the work order is written. The tool designer is to accept and use the tooling description as the title to be placed on the tool drawing.

Tool Numbers

A tool number is an identifying number that will eventually be placed on the tool drawing as the tool number and stamped or etched on the tooling hardware. The tool designer should place a note on the tool drawing requesting that each detail be permanently marked with the tool number. These markings will ensure the traceability of the tooling back to the tool drawing.

Manufacturing companies typically create classifications of tooling with significant numbers (see Chapter 7) to correspond to each tooling classification. These numbers are taken as required from, and recorded in, a log book by the process engineer. Figure 5-14 is an example page from one such log.

Work-Order Dates

The release date, one of three dates listed on the design work order, designates the date that the work order was written and released to the paperwork system. To maintain the accuracy of the release date, the process engineer should be prepared to process the design work order once it has been dated.

The other two dates found on the work order are requested design and build dates. These dates are a declaration of when the design and build of the tooling must be completed, respectively. These dates should be realistic and are necessary, as many work orders are processed at one time. The chief tool designer and toolroom supervisor will use these dates

TOOL RECORD LOG

PART NO. 32246

PART NAME DIFFERENTIAL HOUS. STYLE F

TOOL NO.	NAME / OPER.	MACH. / B. T.	OP. NO. / DEPT.	W.O. / OWNER	DATE / DWG.	QUAN.
F-1	HOLDING FIXT.	CENTRI SPRAY	70	14-2311	11-2-84	4
	SHOT PEEN	2311	14	CO.	D	
F-2	REPAIR FIXT.	LEBLOND	110	14-2351	11-19-84	1
	HUB REPAIR	4730	16	CO.	D	
F-3	DRILL JIG	ALLEN	40	14-2318	12-4-84	1
	DRILL (12) .250 DIA.	4227	28	CO.	D	

Figure 5-14 Tool Record Log

as guideposts when setting departmental priorities or requesting overtime. These individuals should be consulted when the meeting of these dates may be questionable. Finally, the process engineer must consider the work-order approval procedure and its impact on the design and build dates.

Operation Numbers

The operation numbers listed on the design work order have corresponding operation sheets. The design of the tooling being requested must conform to the information set forth on these operation sheets.

Because an operation sheet shows how the part will appear after that operation, the tool designer should also review the operation sheet just prior to the one listed on the work order.

Where more than one operation and/or part number are given on the work order, blueprints of each must be run, for reference purposes, before starting the design process.

Part Numbers

The part numbers listed on the design work order have dual meanings. First, they represent product drawings that can be examined for reference information. Second, they represent operation sheets that were drawn in accordance with the product drawings.

KEY CONCEPT

The tooling requested on the design work order must be designed in total agreement with the operation sheet, while using the product drawing for reference or background information only.

Machine Brass Tag

The machine brass tag is an identification number stamped in brass and affixed to all machine tools and other pieces of capital equipment. Because an individual manufacturing plant may have 10 or more of a particular machine, the brass tag is the only means of positive machine identification.

The corrosion-resistant properties of brass make it an ideal material for in-plant use.

In addition to the obvious need for machine recognition, the brass tag is used to represent all in-plant equipment for accounting, depreciation, and corporate taxation purposes. Therefore, whenever a new piece of capital equipment is to be purchased, the process engineer must acquire a new brass-tag number prior to letting the purchase requisition.

When tooling needs to be designed for a particular piece of equipment, the process engineer will place the appropriate brass-tag (BT) number on the design work order. The tool designer must pay close attention to this BT number and seek out information about that machine before starting the design process.

The Estimate of Cost

The estimate of cost given on the design work order is the process engineer's best approximation of the total dollars required to complete the work requested. This estimate of cost should have been made previously along with other such economic processing considerations (see Chapter 4).

Instructions

A place for specific written instructions is provided on each design work order. In this spot, the process engineer formally communicates with the tool designer. Because the space provided for such instructions is necessarily small, phrases such as those listed below will key the designer to additional information:

1. "See writer"
2. "See reference drawing # ________ "
3. "See attached sketch"

The importance of these written instructions rests on the fact that they represent a formal request; therefore, all pertinent facts, figures, and other information must be listed.

Work-Order Numbers

Before a design work order can be processed, it must be identified with one unique work-order number. Work-order numbers are used to authorize the expenditure of company funds and to accumulate actual charges for labor and material.

In yearly financial planning, manufacturing companies estimate how much money will be required to maintain existing operations, to purchase new equipment, and to cover other specialized expenses. The monitoring of these funds is necessary to keep expenses within the budgeted amounts. Therefore, typical cost-accounting techniques require that work-order numbers be assigned by classification. Each classification is called a work-order series, because a series of numbers have been set aside for use in that class.

Normally, a work-order series will be set up in two specific areas:

1. The "APPROPRIATION" work-order series. These are company funds set apart or assigned for a particular purpose or use (for example, equipment for a new product line).
2. The "EXPENSED" work-order series. These are company funds utilized to maintain normal operations within the plant (for example, repair or replacement of worn-out equipment). These funds are chargeable against company revenues for a specific period of time.

In addition, each work-order series is further broken down to distinguish between capital equipment and durable tooling.

As work-order numbers are taken out by the process engineer, special attention must be given to matching the work requested to the work-order series number being assigned.

Purchase Requisitions

The completed purchase requisition is a formal written request for something to be purchased and can be written by any salaried person within a company (see Fig. 5-15). A purchase order, however, is a legal document (contract) authorizing goods or services to be delivered and paid for. The purchasing department and its buyers and agents are the only individuals in a typical company who are authorized to enter into such agreements.

A work order or a charge number must accompany the purchase requisition to show where the necessary monies are coming from. Although a work order is used to cover the cost of many purchased items, approval signatures from the appropriate management personnel will still be required on the purchase requisition. To minimize the time involved in the approval procedure, the engineer should, whenever possible, complete all necessary

Purchase Requisition

Order From		Ship To	

Account No.	B. T. No.	Dept. No.	Approp. No.	W. O. No.

Via:	F.O.B.	Terms	Deliver To

Wanted	Quantity	Description	Price

To Be Used For		Remarks
Reqd. By	Date	
Approved By	Date	

Figure 5-15 Purchase Requisition

purchase requisitions and attach them to the unapproved work order. (Curtis, *Tool Design,* 1986)

5 ▪ 10 CONCLUSION

At this point, the process engineer should have been given or completed the following ordered "Steps of Process Planning."

1. A request for a cost estimate (CET)
2. A review of product drawings if available
3. A component estimate, including required operations, machines, tooling costs, and production rates
4. Filing of the CET and component estimate for future reference
5. A formal engineering release authorization

6. A review of the CET file and product drawings in light of any possible changes and firm production quantities
7. Determination of a rough lineup of operations (a process sequence)
8. A review of available machines
9. Determination of the percent utilization and production capacity of selected machines
10. Sketching of a rough process in strip-chart form
11. A tolerance chart
12. A firming up of the process sequence and required machines
13. The figuring of all time values (speeds, feeds, production rates, etc.)
14. A confirmation of machine capacity
15. The creation of a formal routing and the required process sheets
16. The listing and sketching of all necessary tooling
17. The writing of work orders and purchase requisitions
18. A final review of all materials
19. The formal release of work orders, purchase requisitions and process sheets
20. An implementation and follow-up plan

Here process planning as treated in this book ends, and process implementation begins. This process implementation will often consist of these additional responsibilities:

1. Tool-design procurement
2. Tool-design approval
3. Tool-build procurement
4. Tool tryout
5. Statistical process analysis
6. Purchase-order follow-up
7. Request for plant layout work (that is, writing a move order)
8. Operator training
9. Shop floor setup and troubleshooting
10. Methods improvement

REVIEW QUESTIONS

1. What is a release and why is it used?
2. What is a tolerance chart used for?
3. How are nondimensional operations documented?
4. What is a tool layout?
5. What is a design work order and who writes it?

CHAPTER 6

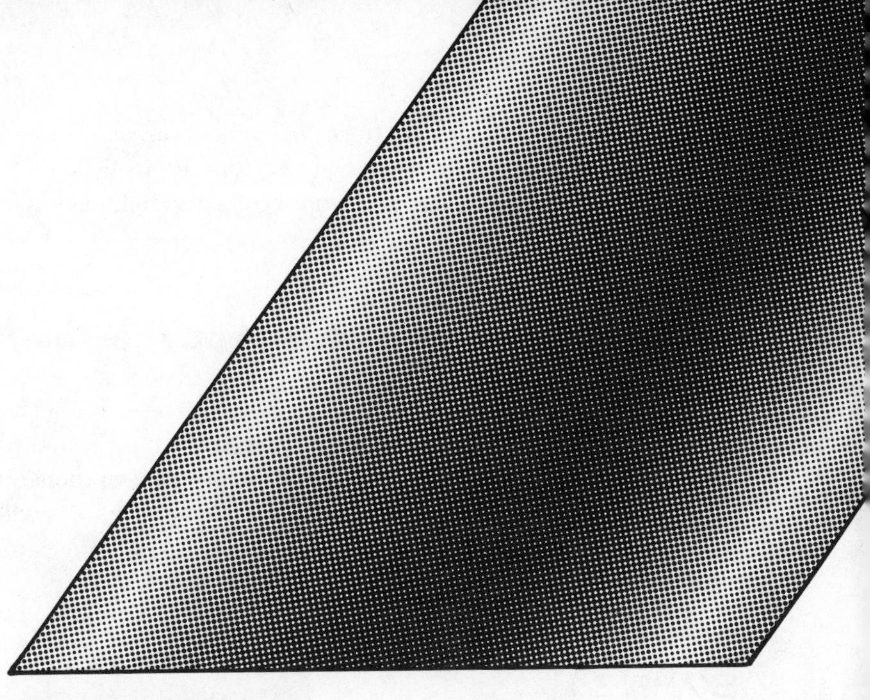

Tolerance Charting

6 ▪ 1 INTRODUCTION

Tolerance charting is a systematic method for establishing proper in-process dimensions and tolerances of the type required on individual process sheets. Simple economics dictate that all process tolerances be maximized through correct process sequencing and datum selection. Tolerance charting provides the process engineer with a means of rapidly comparing a number of processing options. It also furnishes the process engineer with the documentation necessary to confidently arrive at correct process dimensions and tolerances.

Tolerance charting is specifically designed for use in the creation of length dimensions and tolerances of the type typically found on discrete parts that are processed through a variety of conventional machining processes.

Product Drawing versus Process Tolerances

Before tolerance charting can be explained, the need for such a technique must be understood. When tolerances given on the product drawing are compared with those dictated by processing considerations, the value of tolerance charting should become apparent. Two examples will be used to demonstrate this concept.

EXAMPLE 1

In this example, the process engineer has received a product drawing of a simple detail containing three holes (see Fig. 6-1). It is assumed that the location of these holes have been toleranced in accordance with the functional requirements of the workpiece. This means that all process tolerances must be equal to or better than the product-drawing tolerances.

Further, it is assumed that due to the production volume required, the process engineer decided to drill all three holes simultaneously. This means that datums *A* and *B* and the three holes had to be located with a multiple spindle drill head and drill jig. A process sheet depicting this operation is shown in Figure 6-2. Note that

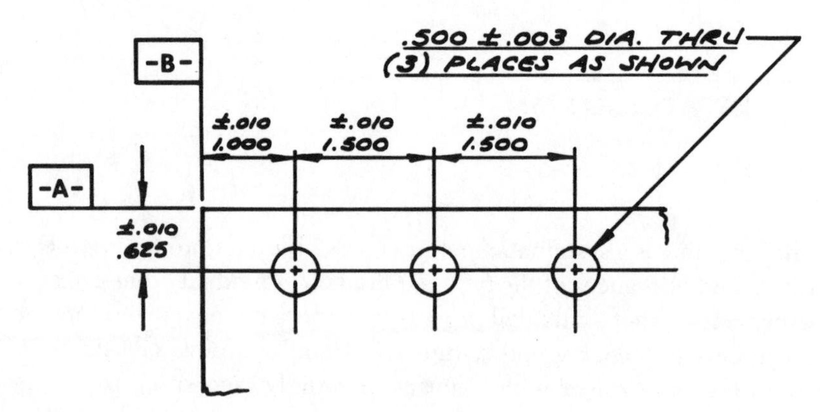

Figure 6-1 Sample Product Drawing

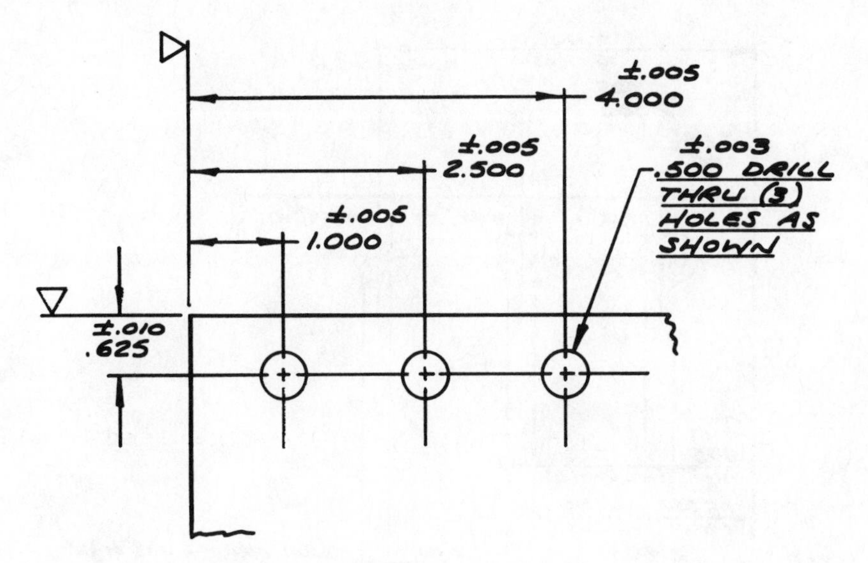

Figure 6-2 Sample Process Drawing

the dimensions and tolerances listed on the process sheet do not match those listed on the product drawing. Why?

Upon closer examination, the necessity of these seemingly tighter process tolerances will become evident (see Fig. 6-3). Comparison of the minimum dimensional possibilities of each hole shows that the resultant dimension of 1.510/1.490 between holes matches perfectly the dimension and tolerance shown on the product drawing.

EXAMPLE 2

The process engineer has received a product drawing of a detail containing a series of counterbored holes (see Fig. 6-4). The product designer has dimensioned the depth of the counterbore from surface *A*. From this information it can be assumed that the maximum head height of the screw to be placed in the counterbore is shorter than .385 inches. This will ensure that the head of the screw will not protrude above surface *A*.

In processing this part, the process engineer decided to clamp the part against surface *B* and dimension the depth of the counterbore from this surface (see Fig. 6-5). Again, note how the dimensions and tolerances listed on the process sheet do not match those listed on the product drawing. Why?

Figure 6-6 serves to show how the permissible variation in the 1.000 dimension works together with the relative position of the counterbore cutter to surface *B* in generating the actual depth of the counterbore from surface *A*. The relative position

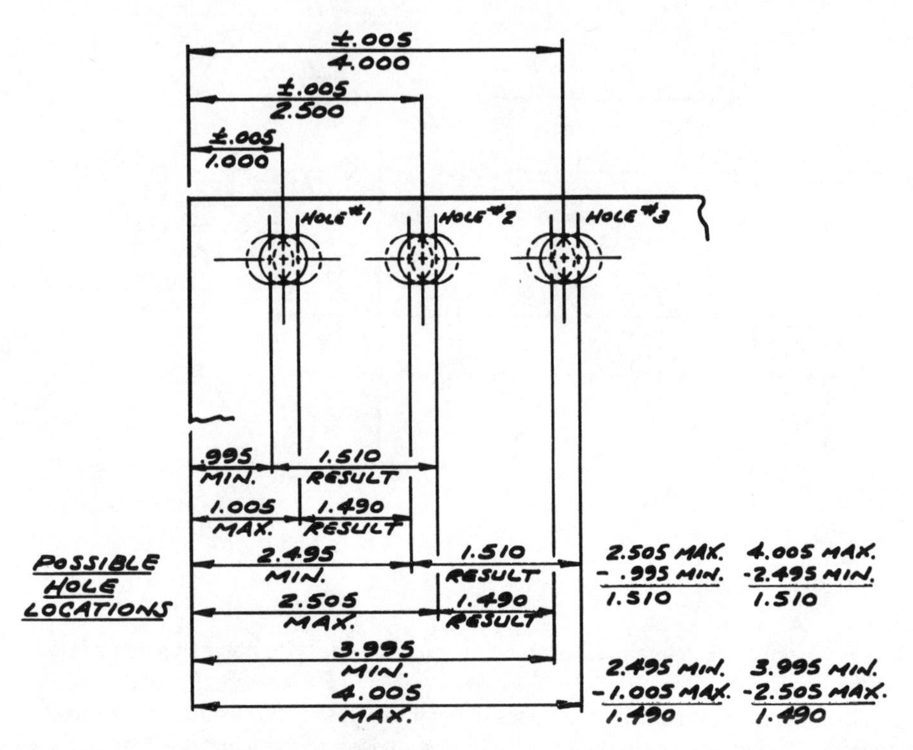

Figure 6-3 Possible Hole Locations

of the counterbore cutter to the locating surface is all that can be controlled at this operation. Therefore, it is the .600-in. dimension that must be toleranced to ensure dimensional compliance with the product drawing.

Summary of Tolerance Considerations

From the preceding examples it can be seen that the way in which a part is processed impacts how it must be dimensioned and toleranced. Figures 6-1 to 6-6 illustrate the kind of thinking process engineers have used to correctly process their production parts. The sketching of minimum and maximum dimensional conditions, the subtractions, and the eventual common-sense double checks are fine for simple processing exercises. However, when a dozen or more processes are involved, the old minimum/maximum method of tolerancing process sheets becomes too complicated to manage. That is when the technique of tolerance charting can be justified, explained, and understood.

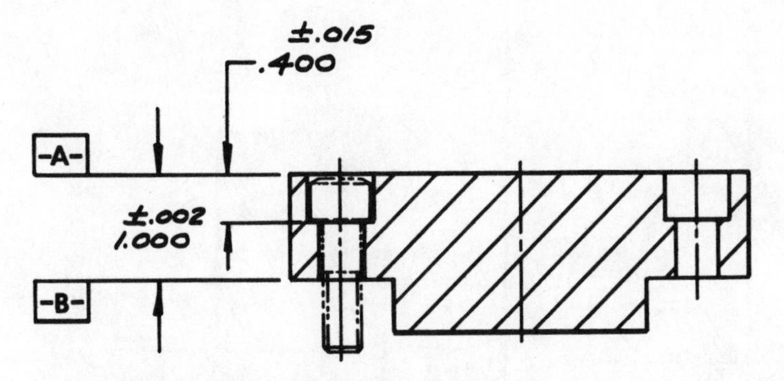

Figure 6-4 Sample Product Drawing

This chapter is designed to systematically walk the engineer through the basic steps of tolerance charting. Upon completion, the tolerance chart becomes a semipermanent record of how all process dimensions for a given part were developed.

6 ▪ 2 PRELIMINARY STEPS

Before the tolerance chart can be created, a series of preliminary decisions must be made. These decisions are characterized by the following four steps and their related documents.

Step I—Simplified Tolerance Charting

Once the process engineer has been given a product drawing to process (see Fig. 6-7), she or he is faced with the problem of identifying critical locating

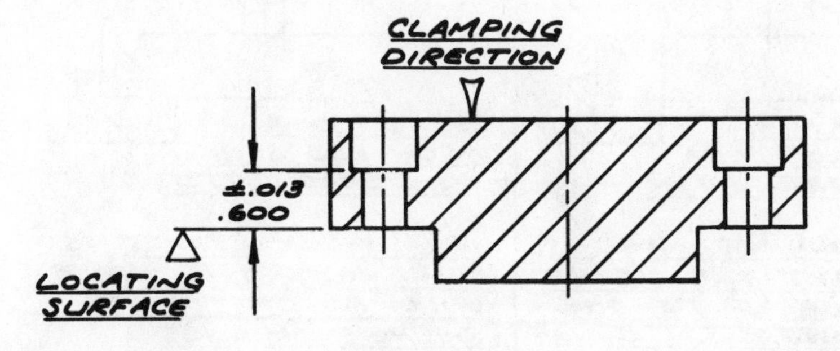

Figure 6-5 Sample Process Drawing

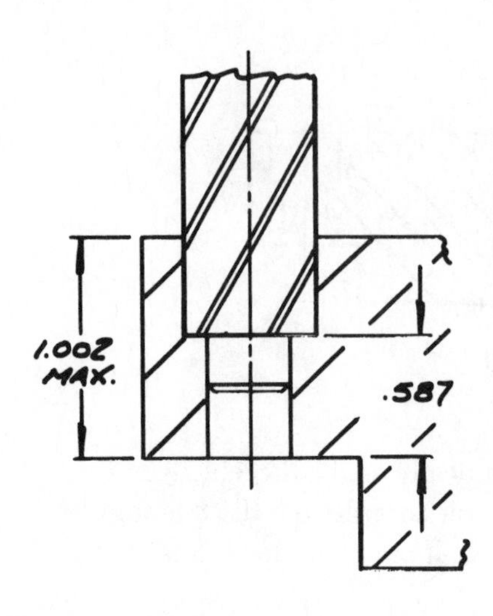

1.002 MAX.
−.587 MAX. DEPTH
.415

1.002 MAX.
−.613 MIN. DEPTH
.389

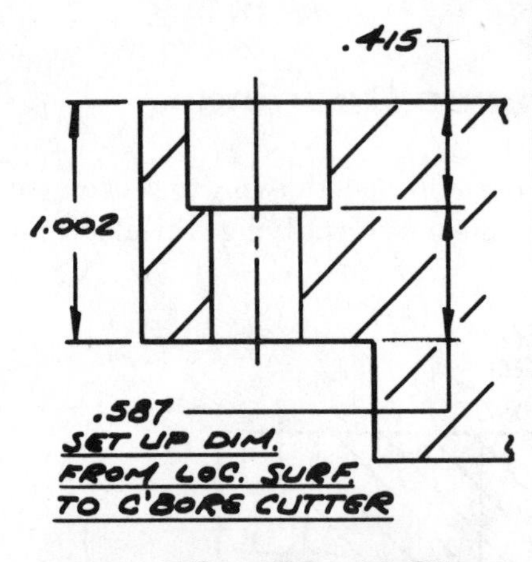

.587
SET UP DIM.
FROM LOC. SURF.
TO C'BORE CUTTER

IF A THINNER PART IS
RUN NEXT, THE C'BORE
WILL BE UP TO .004
SHALLOWER OR .411

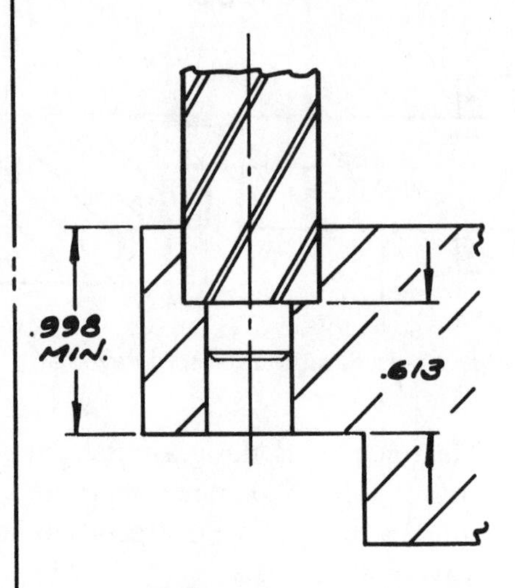

.998 MIN.
−.613 MIN. DEPTH
.385

.998 MIN.
−.587 MAX. DEPTH
.411

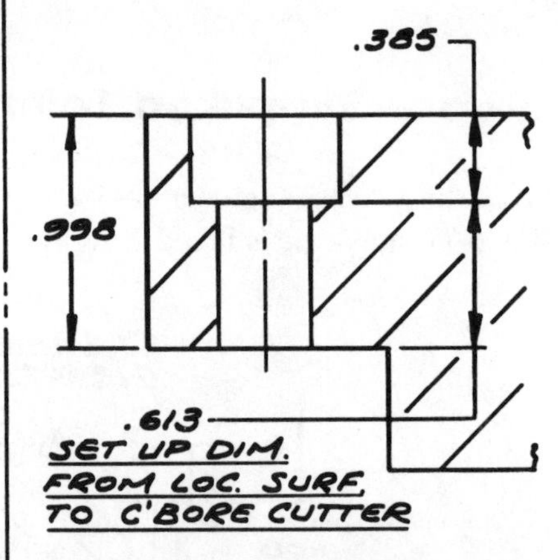

.613
SET UP DIM.
FROM LOC. SURF.
TO C'BORE CUTTER

IF A THICKER PART IS
RUN NEXT, THE C'BORE
WILL BE UP TO .004
DEEPER OR .389

Figure 6-6 Process Possibilities

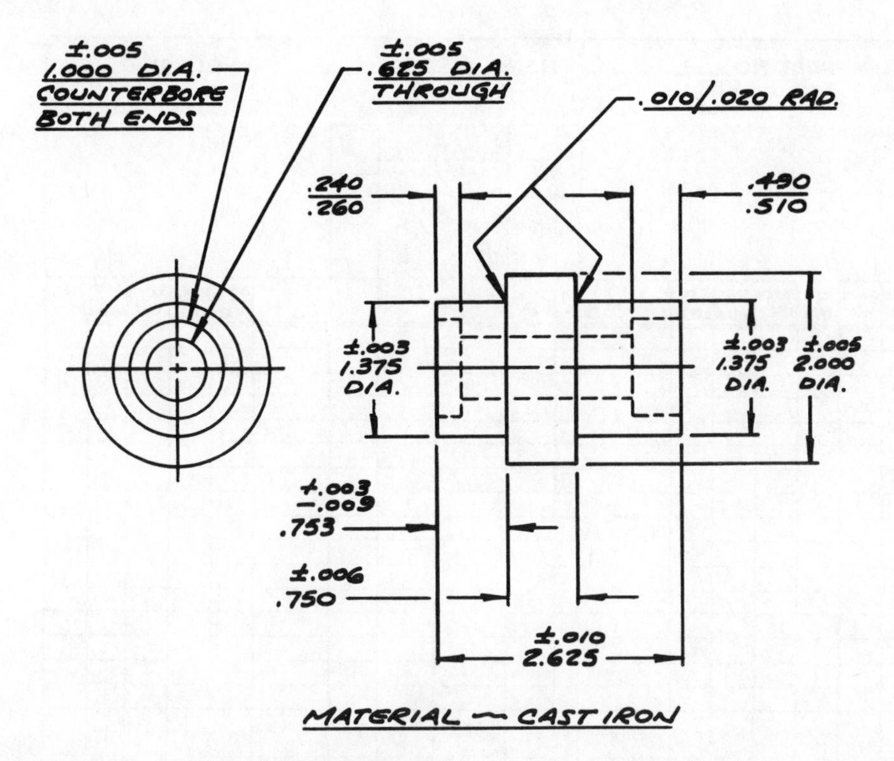

Figure 6-7 Sample Process Drawing

surfaces prior to establishing a sequence of operations. The creation of a simplified tolerance chart (see Fig. 6-8) can aid in this datum identification. This concept was previously discussed in Chapter 2, Section 2.4.

Note in Figure 6-8 that a half-sectional view of the part is drawn at the top of the chart. Then, extension lines are drawn from each length surface to the bottom of the chart.

At this point, lines representing the length of each dimension on the product drawing are placed between the chart's extension lines in the area below the line labeled *blueprint*. The lines are called balance dimensions and have a dot placed at each end of each line.

Adjacent to the balance dimensions, mean dimensional values and equal bilateral tolerances for each product-drawing dimension must be penciled in. This step is required for later manipulation and splitting of tolerances between two or more cuts.

Now, by simply counting the number of dots resting on each surface (extension line), the process planner is able to see the most important surfaces.

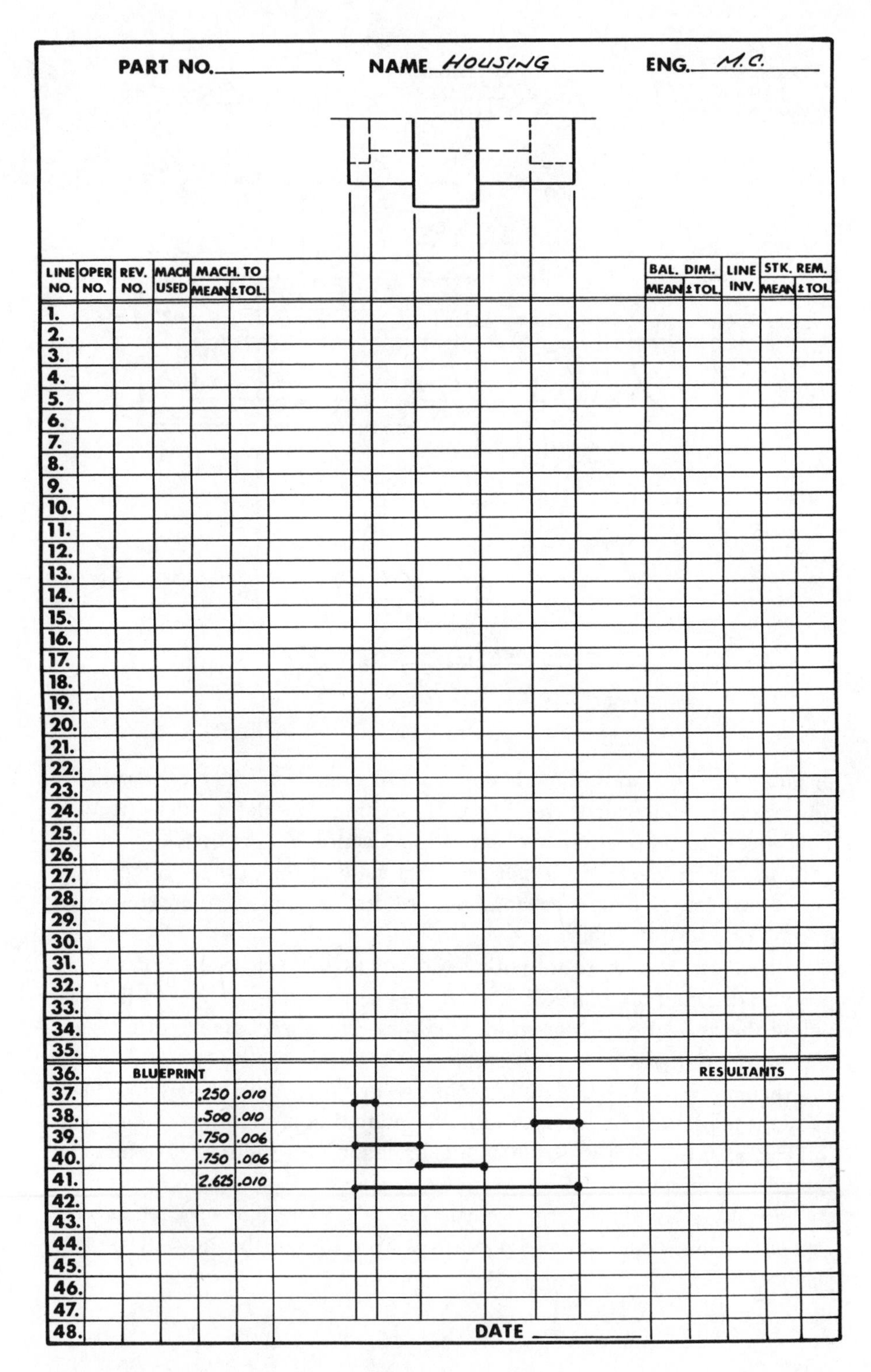

Figure 6-8 Simplified Tolerance Chart

Step II—Sequences of Operations

With a firm understanding of the critical dimensions and locating surfaces that have been identified by way of the simplified tolerance chart, the process planner can develop a tentative sequence of operations or routing (see Fig. 6-9).

Step III—Machine Selection

As the sequence of operations is developed, machines for each operation should also be selected. Knowledge of each machine's dimensional capabilities will be necessary when decisions about splitting tolerances between operations are later required.

NAME SAMPLE DATE ______ PART NO. ______
ASSY. NO. ______ CUST. ______
MATL. ______ SHEET NO. ______

Dept	Description	Sheet	Oper	Hourly Production	Rate	Labor	Set-Up Hours
	PURCHASE CASTING		—				
	MACHINE RIGHT HAND END OF CASTING AND DRILL THRU HOLE (MACHINE ~ CNC LATHE)		10				
	DURING SECOND CHUCKING MACHINE LEFT HAND END OF CASTING (MACHINE ~ CNC LATHE)		20				

Date	Change	By	Date	Change	By	Date	Change	By

Figure 6-9 Routing, Tentative Sequence of Operations

Step IV—Strip Chart

A strip chart is a series of sketches laid out on a strip of paper showing how the part will tentatively be processed (see Fig. 6-10). It can be thought of as a series of crude process sheets.

Of course, at this point, the engineer has no idea what the actual dimensions and tolerances will end up being. However, based on the selection of each operational locating surface, the engineer does know how each individual process must be dimensioned. Therefore, blank dimensional lines are shown on the strip layout at this point. Later, these dimensional lines will be transferred to the tolerance chart. Still later, dimensions and tolerances arrived at on the tolerance chart will be placed back on the strip layout.

6 ▪ 3 CHART PREPARATION

Having arrived at a tentative sequence of operations, including all clamping- and locating-points, specific machining dimensions, and the basic state of the raw material, the process planner is ready to begin formal preparation of the tolerance chart. The chart preparation is broken down into six steps as listed below. These steps are also graphically illustrated in Figures 6-11 to 6-18.

Chart Preparation Steps

1. Draw in a half-view of the part at the top of the chart. This drawing should also include an indication of the raw material to be removed (see Fig. 6-11).
2. Extend all plane lines from the part to the bottom of the chart in the form of extension lines (see Fig. 6-12).
3. Add all blueprint dimensions (product-drawing dimensions) as balance dimensions at the bottom of the chart. Also, convert all product-drawing dimensions to mean dimensions with equal bilateral tolerances, and pencil them in adjacent to their corresponding balance dimensions (see Fig. 6-13). Note that steps 1–3 were previously used in the creation of a simplified tolerance chart.
4. Enter cut lines, in the order in which they take place, starting at the top of the chart and working downward. Skip 4 to 5 lines between

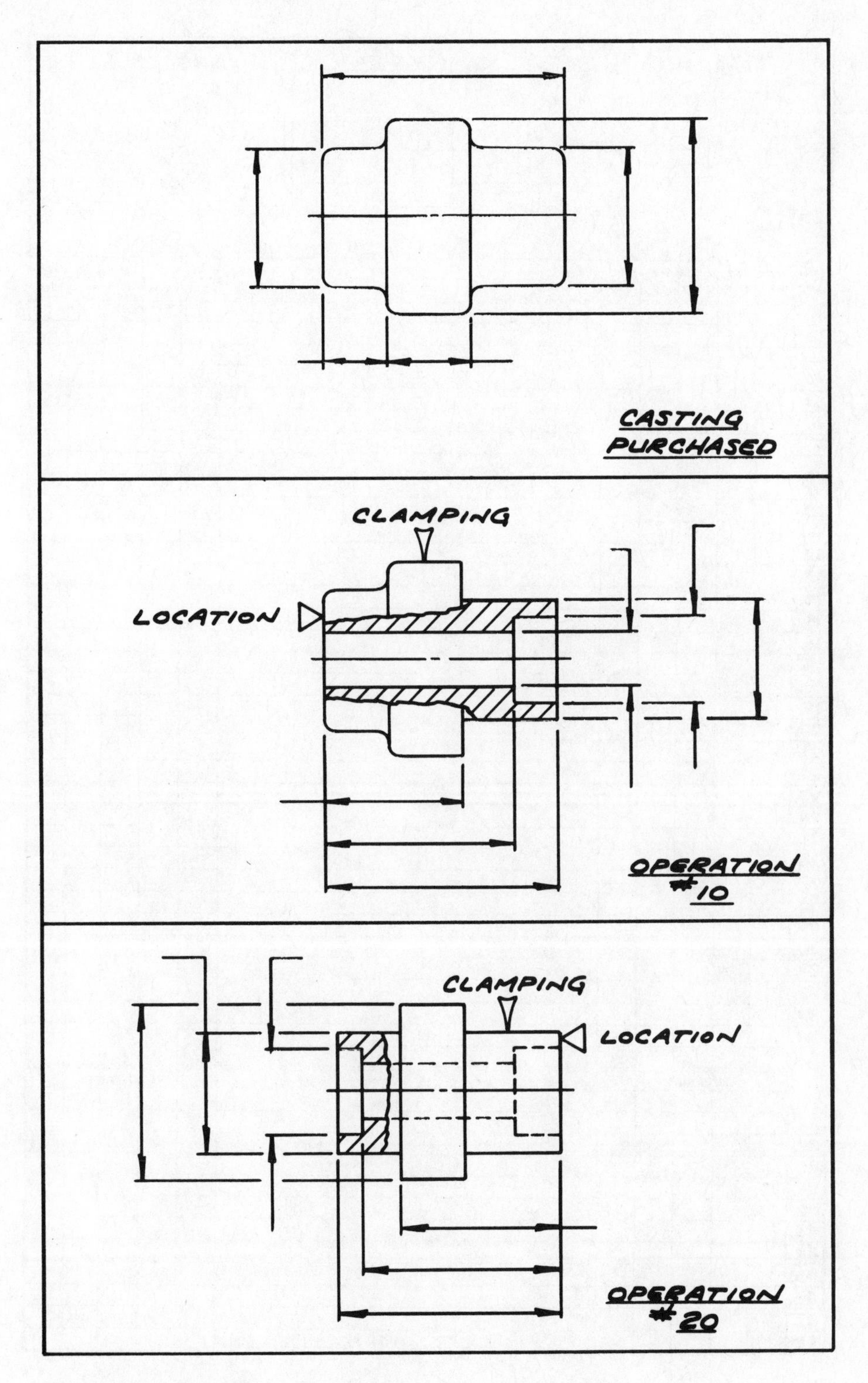

Figure 6-10 Strip Chart

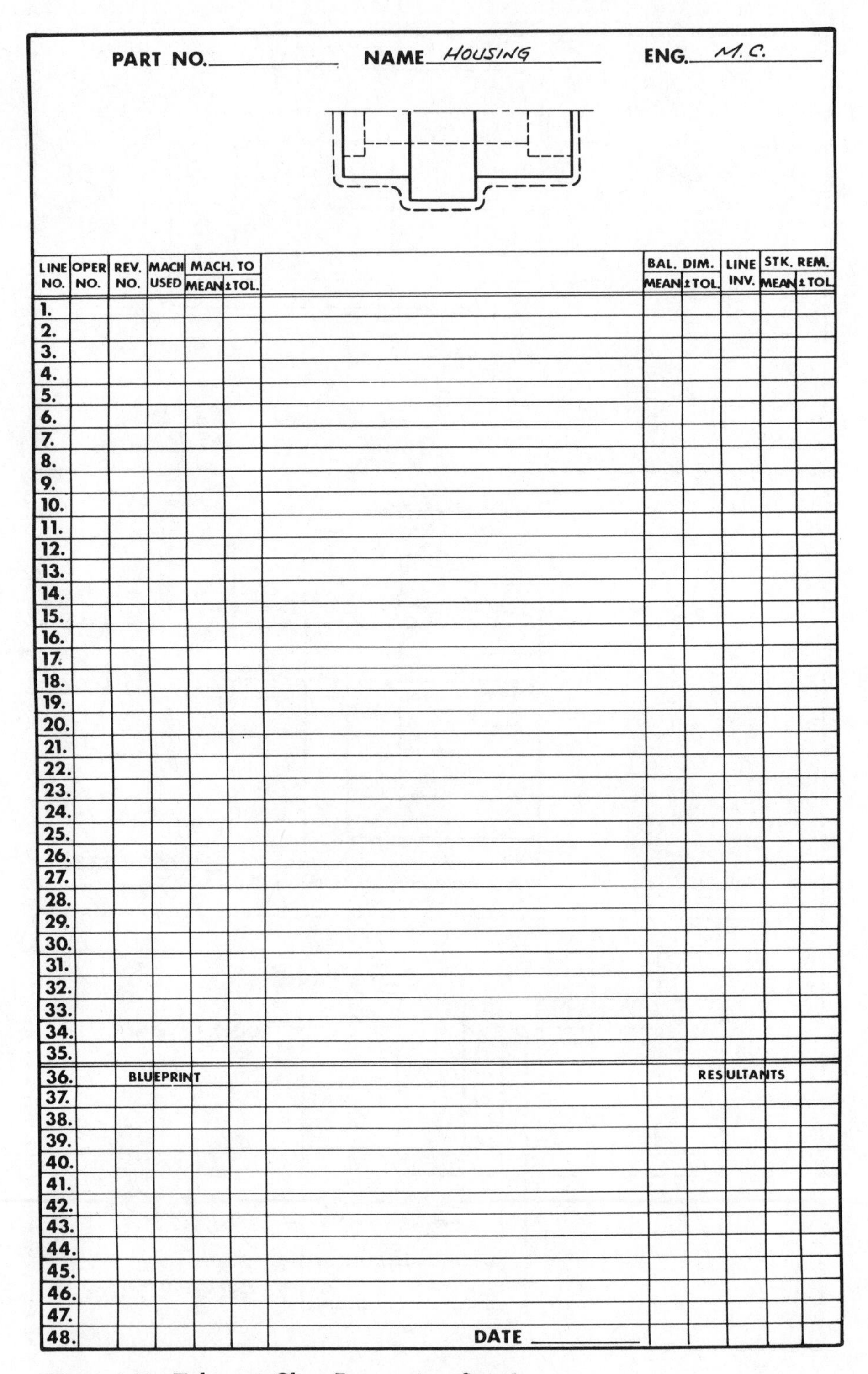

PART NO.________ NAME *HOUSING* ENG. *M.C.*

LINE NO.	OPER NO.	REV. NO.	MACH USED	MACH. TO MEAN	±TOL.		BAL. DIM. MEAN	±TOL.	LINE INV.	STK. REM. MEAN	±TOL.
1.											
2.											
3.											
4.											
5.											
6.											
7.											
8.											
9.											
10.											
11.											
12.											
13.											
14.											
15.											
16.											
17.											
18.											
19.											
20.											
21.											
22.											
23.											
24.											
25.											
26.											
27.											
28.											
29.											
30.											
31.											
32.											
33.											
34.											
35.											
36.		BLUEPRINT							RESULTANTS		
37.											
38.											
39.											
40.											
41.											
42.											
43.											
44.											
45.											
46.											
47.											
48.						DATE ________					

Figure 6-11 Tolerance Chart Preparation, Step 1

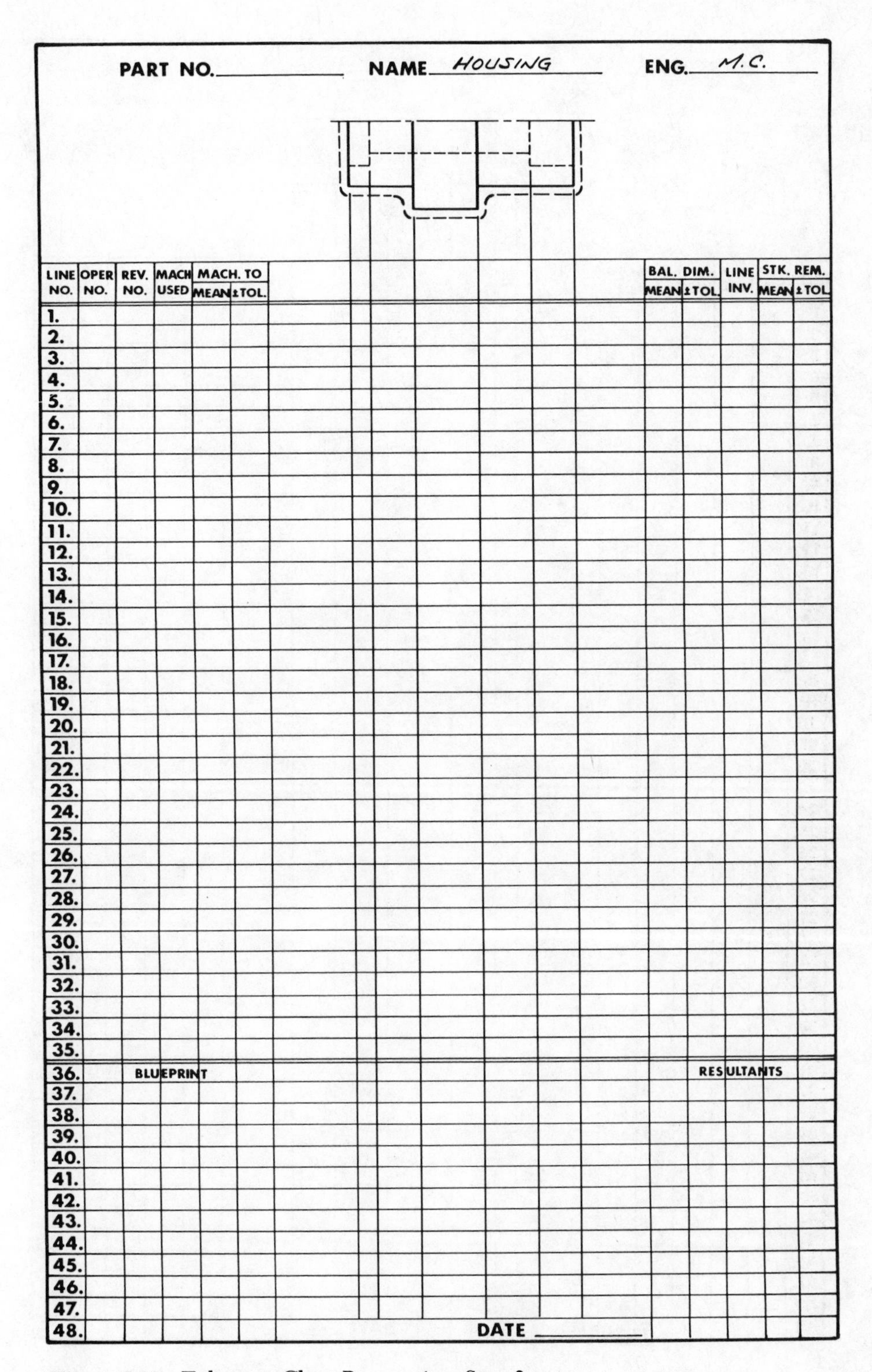

Figure 6-12 Tolerance Chart Preparation, Step 2

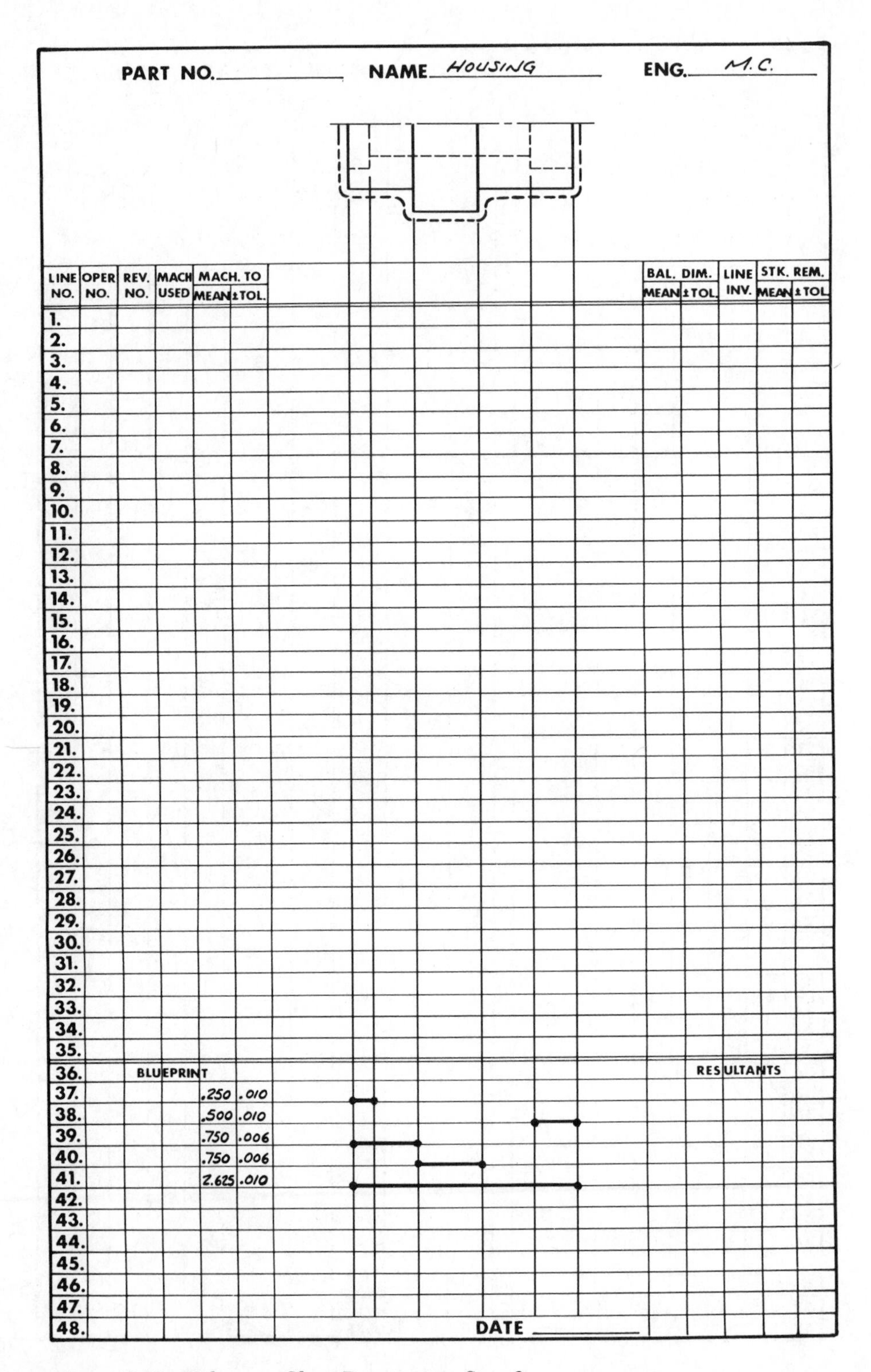

Figure 6-13 Tolerance Chart Preparation, Step 3

each cut line. A cut line has an arrow at one end and a dot at the other. The arrow points at and touches the surface being cut. The dot is an implied location point (see Fig. 6-14).

5. In the appropriate columns next to each cut line, enter corresponding operation numbers and machines used (see Fig. 6-15).
6. Angle in stock to be removed to the line/arrow in which the stock is actually cut off (see Fig. 6-16).

6 ▪ 4 COMPUTING DIMENSIONS AND TOLERANCES

The process engineer is now ready to use the tolerance chart to actually compute final process dimensions and tolerances. This final phase of tolerance charting is complex and involves 10 specific steps that must be rigidly followed.

Dimension-and-Tolerance Computation Steps

STEP 1

Create a series of machining balance dimensions by sliding the blueprint balance dimensions up until each one hits a cut-line arrow. Then, number each balance dimension, and pencil in the line number that affects the tolerance of each balance dimension (see Fig. 6-17).

It can be seen in Figure 6-17 that the .250 ± .010 blueprint dimension was slid up and became balance-dimension line number 31. Further, it can be seen that cut lines 30 and 26 were involved in producing line 31.

Next, the .500 ± .010 blueprint dimension is taken up to become balance dimension line 19. Again, it can be seen that cut lines 18 and 14 work together to produce line 19.

This logic is further employed to create balance dimension number 32, which is brought about through cut lines 30 and 23.

In some cases, balance dimensions that do not appear as blueprint dimensions must be created to properly figure final machining tolerances. One such case is shown as the .750 ± .006 blueprint dimension is brought up to become balance-dimension number 23. Here, as one looks above line 23, it can be seen that cut line 22 affects line 23 as do cut lines 18 and 10. However, in tolerance charting, each balance dimension must be able to be traced to two and only two lines. This rule dictated the creation of

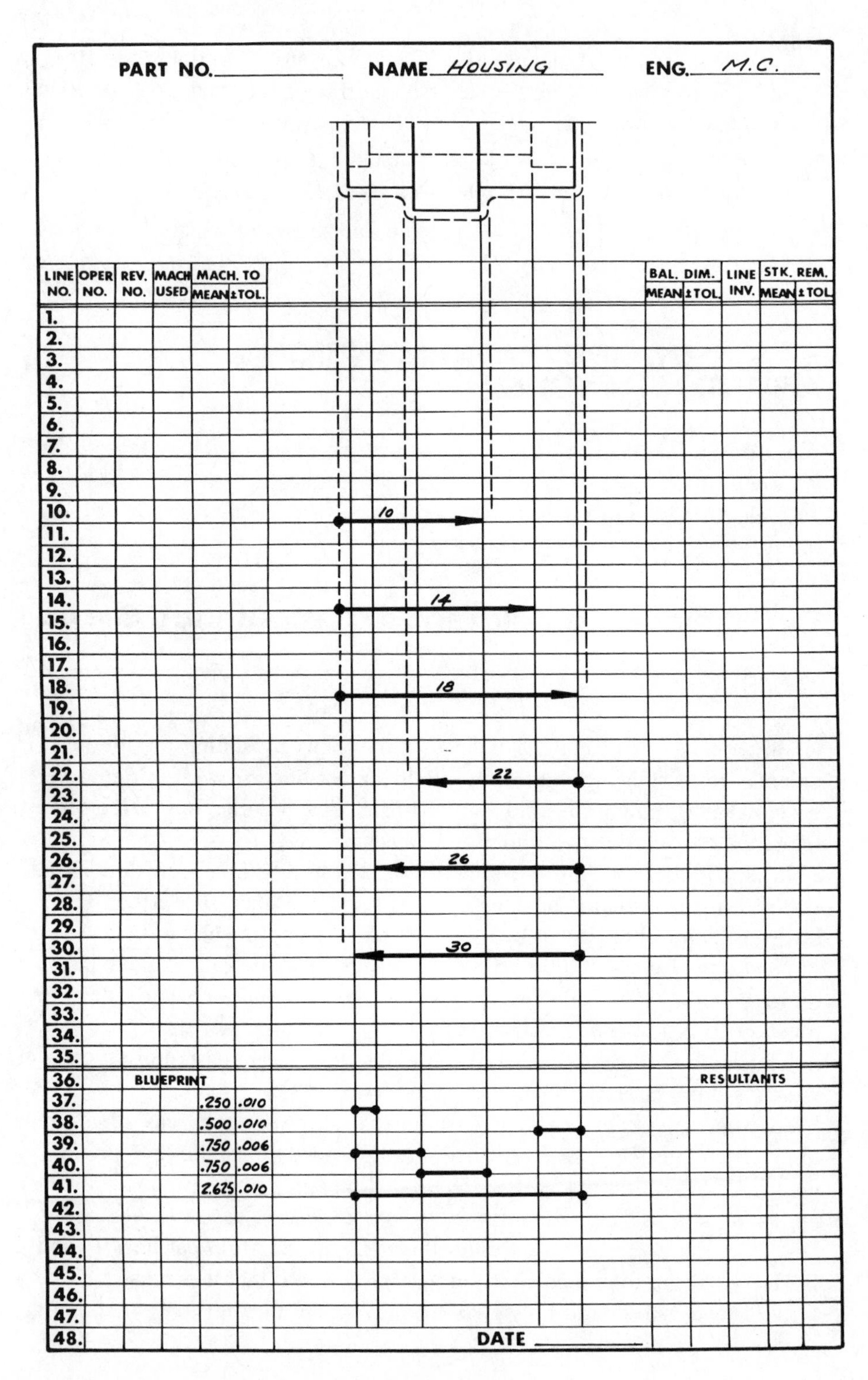

Figure 6-14 Tolerance Chart Preparation, Step 4

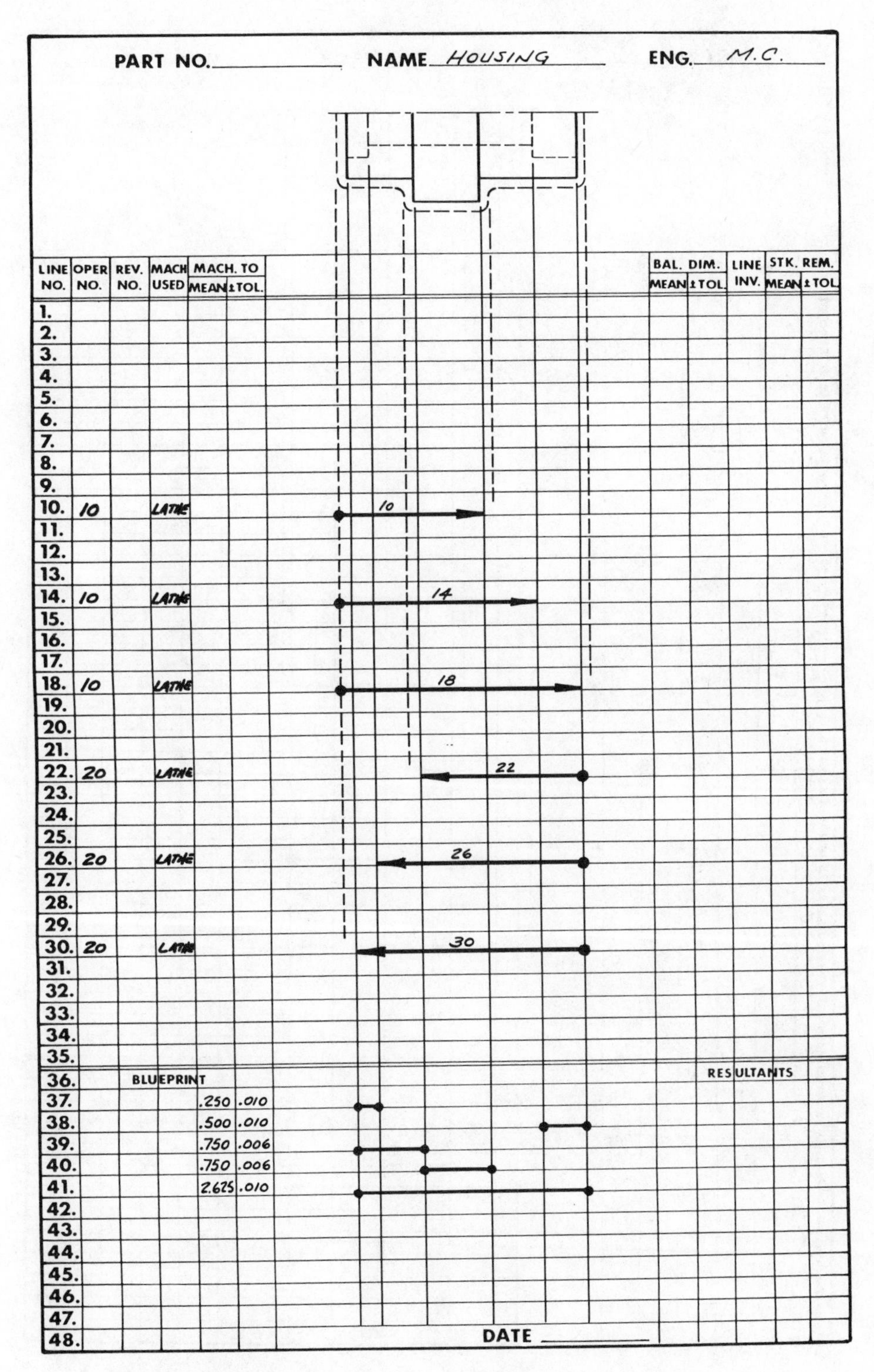

Figure 6-15 Tolerance Chart Preparation, Step 5

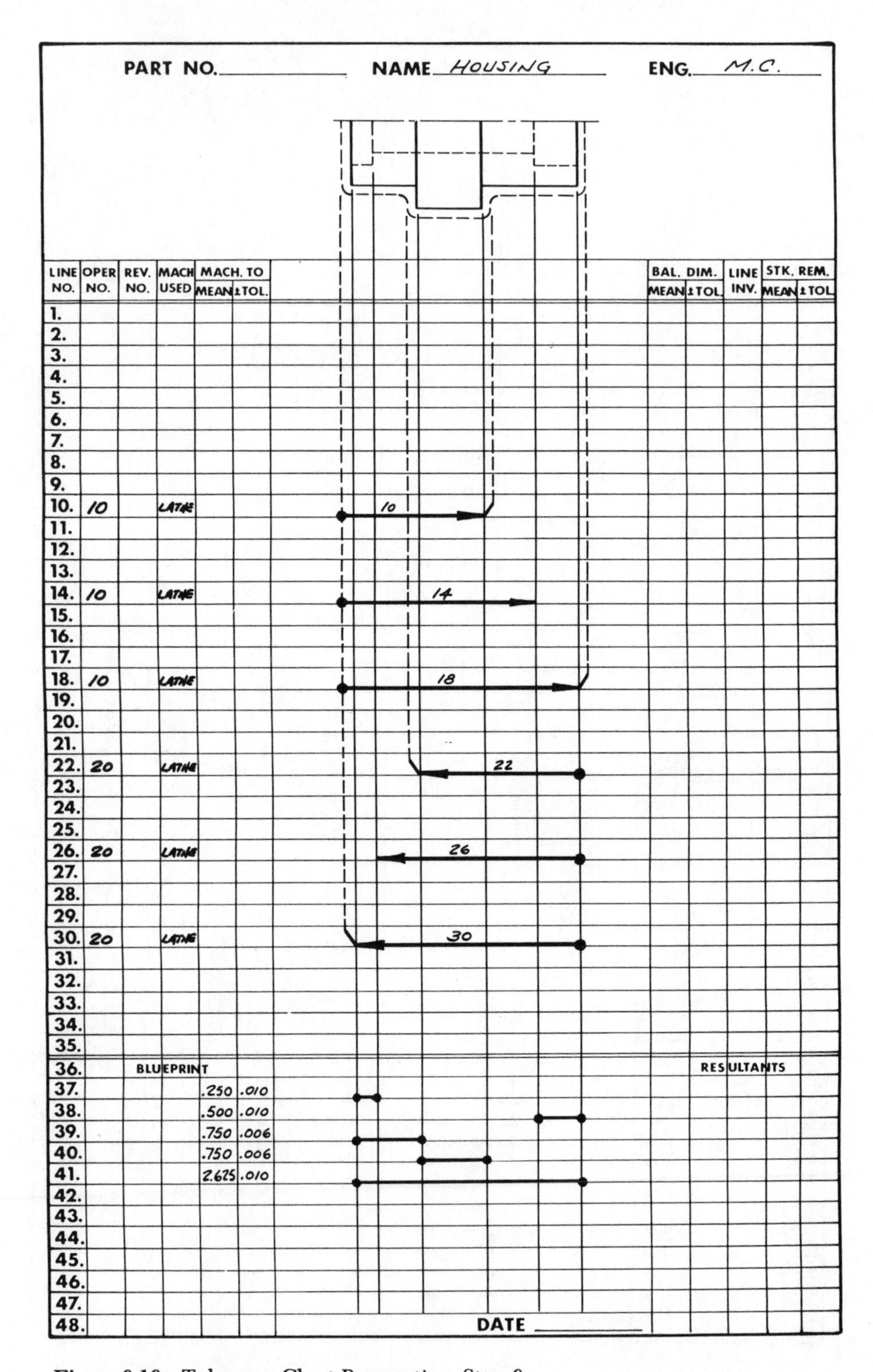

Figure 6-16 Tolerance Chart Preparation, Step 6

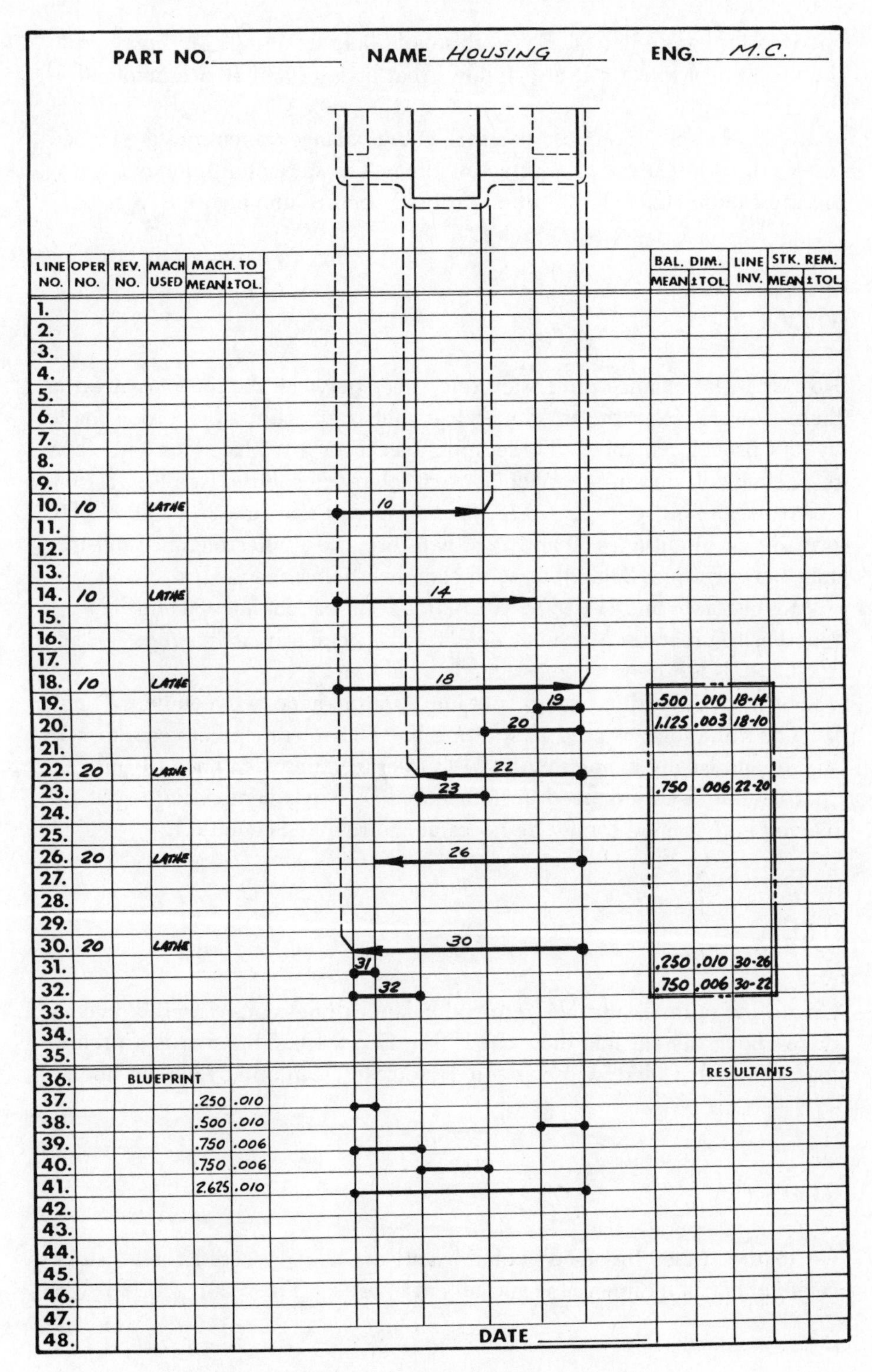

Figure 6-17 Dimension-and-Tolerance Computation, Step 1

balance-dimension line 20. So, in this case, lines 22 and 20 are involved in the creation of line 23. It also follows that lines 18 and 10 are involved in the creation of line 20.

It should also be noted that all machining balance dimensions are placed or found below the cut lines affecting them. The student of tolerance charting must understand this all-important step before moving on to Step 2.

STEP 2

Now split the balance-dimension tolerances between the lines involved in their creation. How the tolerances are split is an arbitrary decision made by the process engineer. These split tolerances are placed next to their respective cut lines in the machine-to ± tolerance column (see Fig. 6-18).

Again, looking at Figure 6-18, it can be seen that line 32 (.750 ± .006) is made up of cut lines 30 and 22. Therefore, the ± .006 tolerance must be split between lines 30 and 22, with each line being assigned ± .003 in.

Moving on to line 31 (.250 ± .010), it is seen that cut lines 30 and 26 work together to create it. Since cut line 30 was previously assigned ± .003 in., ± .007 in. is left over to be assigned to cut line 26.

This logic is used in turn to split line 23's tolerance between lines 22 and 20. The same logic is used in splitting line 20's tolerance between lines 18 and 10 and finally in splitting line 19's tolerance between lines 18 and 14.

If trouble is experienced in understanding why tolerances are split between two cut lines, it may be necessary to review Section 6.1.

STEP 3

Create and number all stock-removal balance dimensions required, placing them above the cut line they affect (see Fig. 6-19). This step is a prelude to ensuring that there will be enough stock to clean up at each cut line.

STEP 4

Go to the "Lines Involved" column and list lines involved for all stock-removal balance dimensions and all cut lines (see Fig. 6-20).

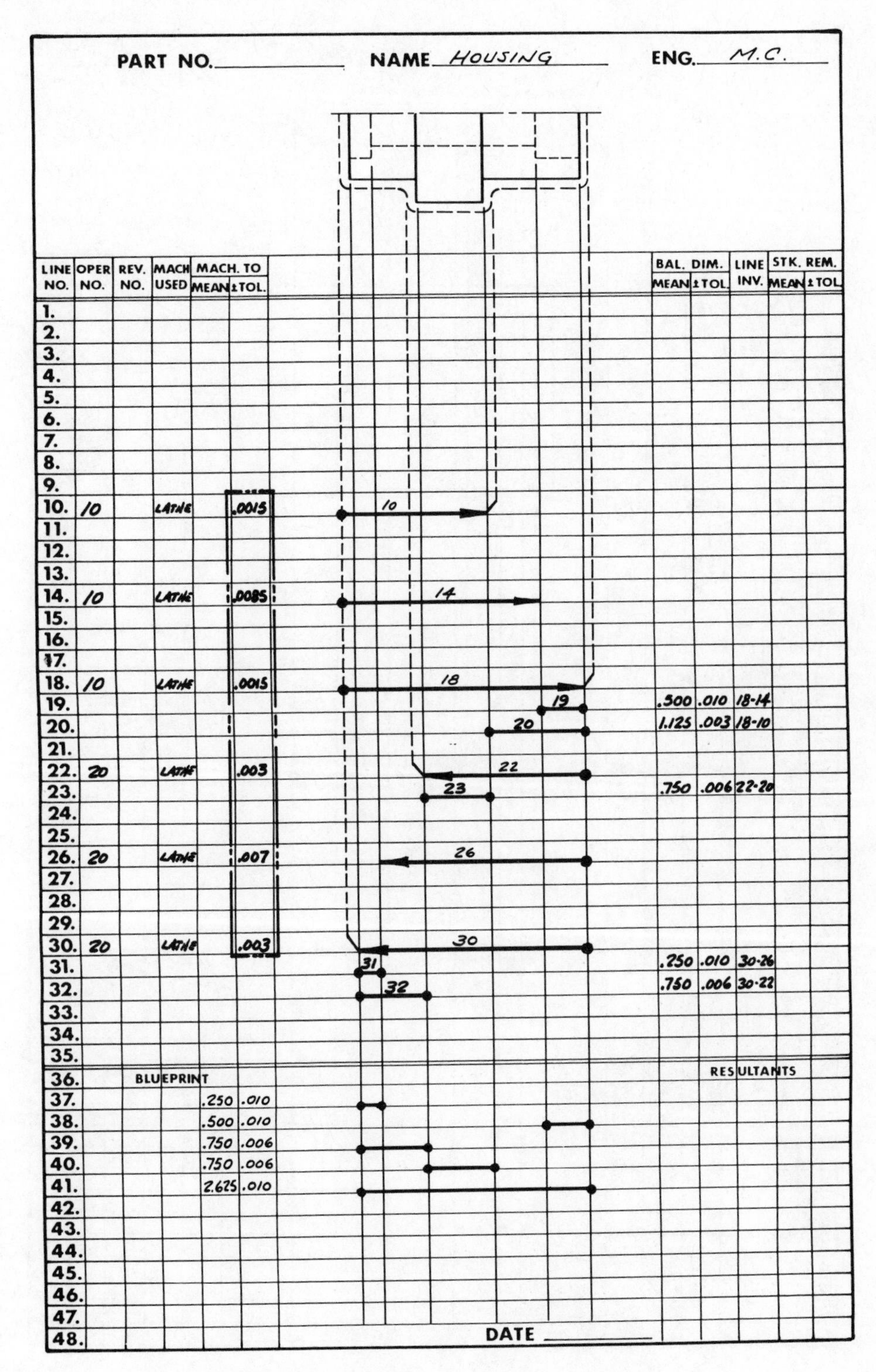

Figure 6-18 Dimension-and-Tolerance Computation, Step 2

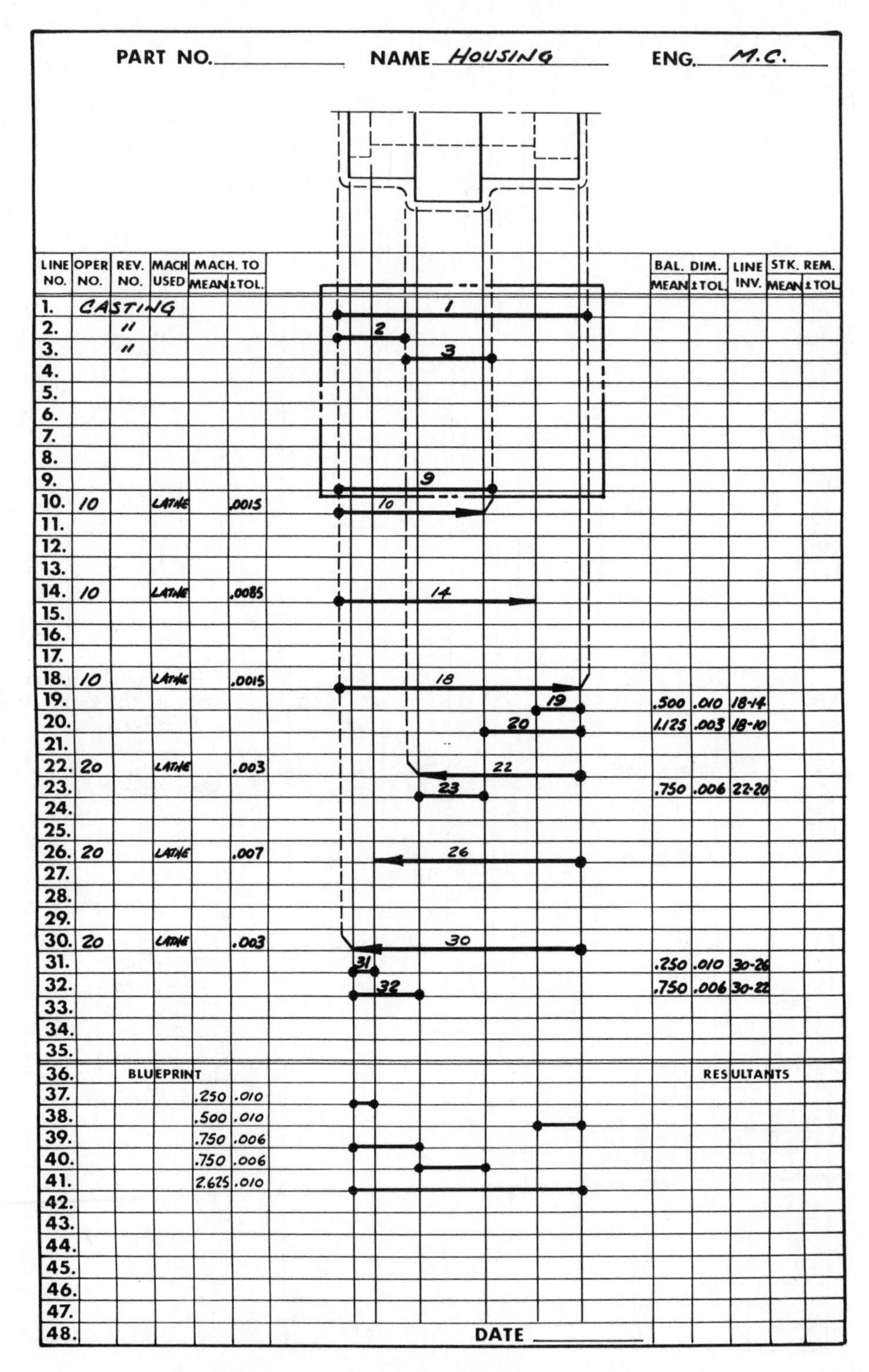

Figure 6-19 Dimension-and-Tolerance Computation, Step 3

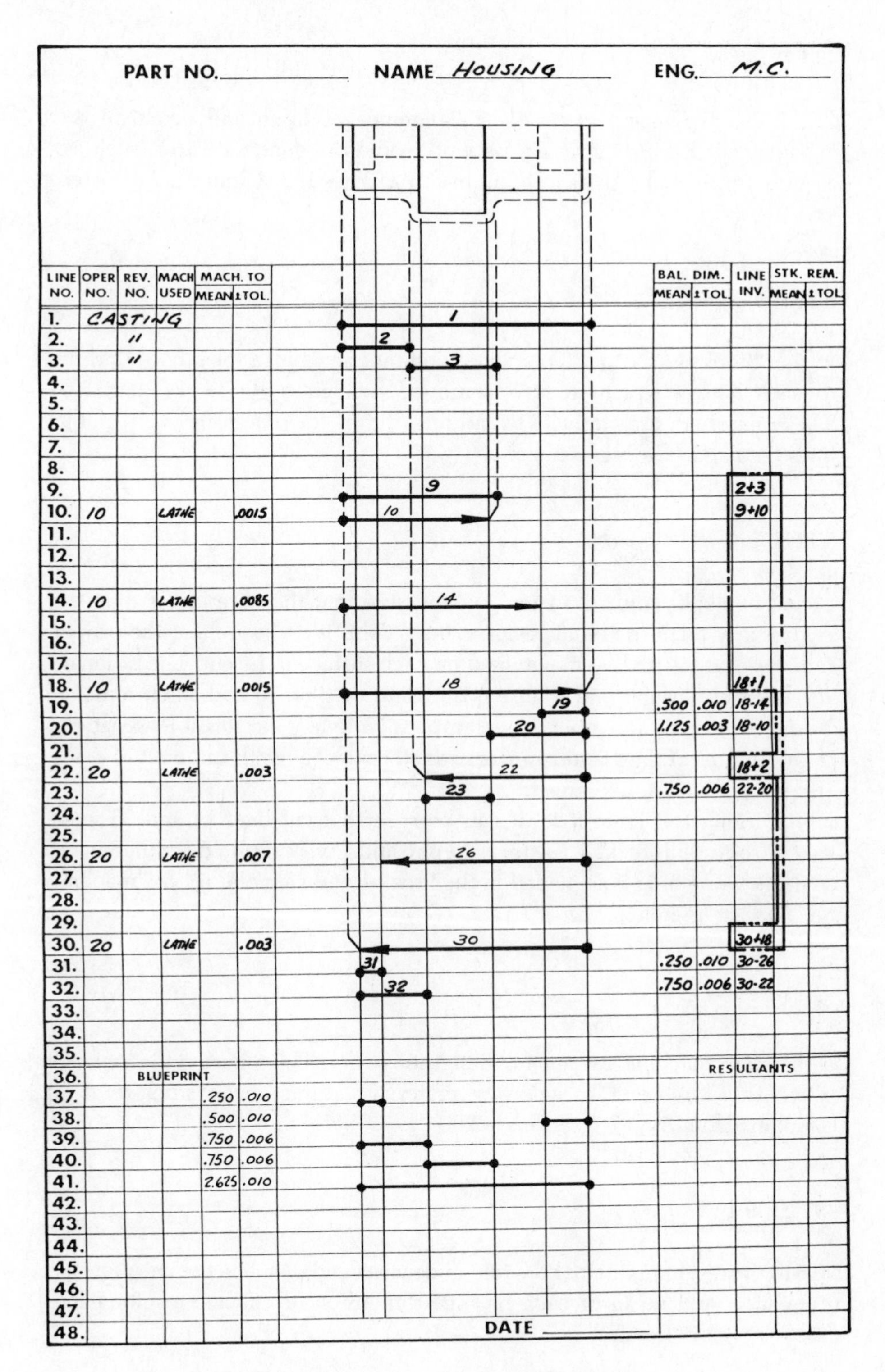

Figure 6-20 Dimension-and-Tolerance Computation, Step 4

STEP 5

Go to the "Balance Dimension ± Tolerance" column and pencil in the ± tolerance that can normally be held in the raw-material form (such as, casting, forging, and so on). In this example, lines 1, 2, 3, and 9 are affected (see Fig. 6-21).

STEP 6

In the "Stock Removal ± Tolerance" column, pencil in a tolerance adjacent to each line in which stock is removed (i.e., lines 10, 18, 22, and 30). These numbers are derived by adding the ± tolerances of each of the lines involved (see Fig. 6-22).

STEP 7

Pencil in stock-removal means, making sure that they are larger than the figures arrived at in Step 6 (see Fig. 6-23). This is to ensure that the proper amount of stock will be maintained on each surface to be cut. The rationale for this step is shown in Figure 6-24, using cut line 10 and stock-removal balance-dimension number 9 as examples. Note how permissible variations in the lengths of lines 9 and 10 work together to change the amount of stock that may be left for removal.

Returning to Figure 6-23, it can be seen that cut lines 14 and 26 were taken from solid stock. Therefore, Steps 6 and 7 were omitted for these cuts, and the word *solid* was placed in the Stock Removal columns of "machine to" and "± tolerance."

STEP 8

Pencil in "machine to" mean dimensions for all cut lines and "balance dimension" means for each raw material, adding mean stock-removal amounts from Step 7 (see Fig. 6-25).

STEP 9

Go to the resultants area of the tolerance chart and pencil in the line number, tolerance, and mean of each balance dimension or cut line resulting in a

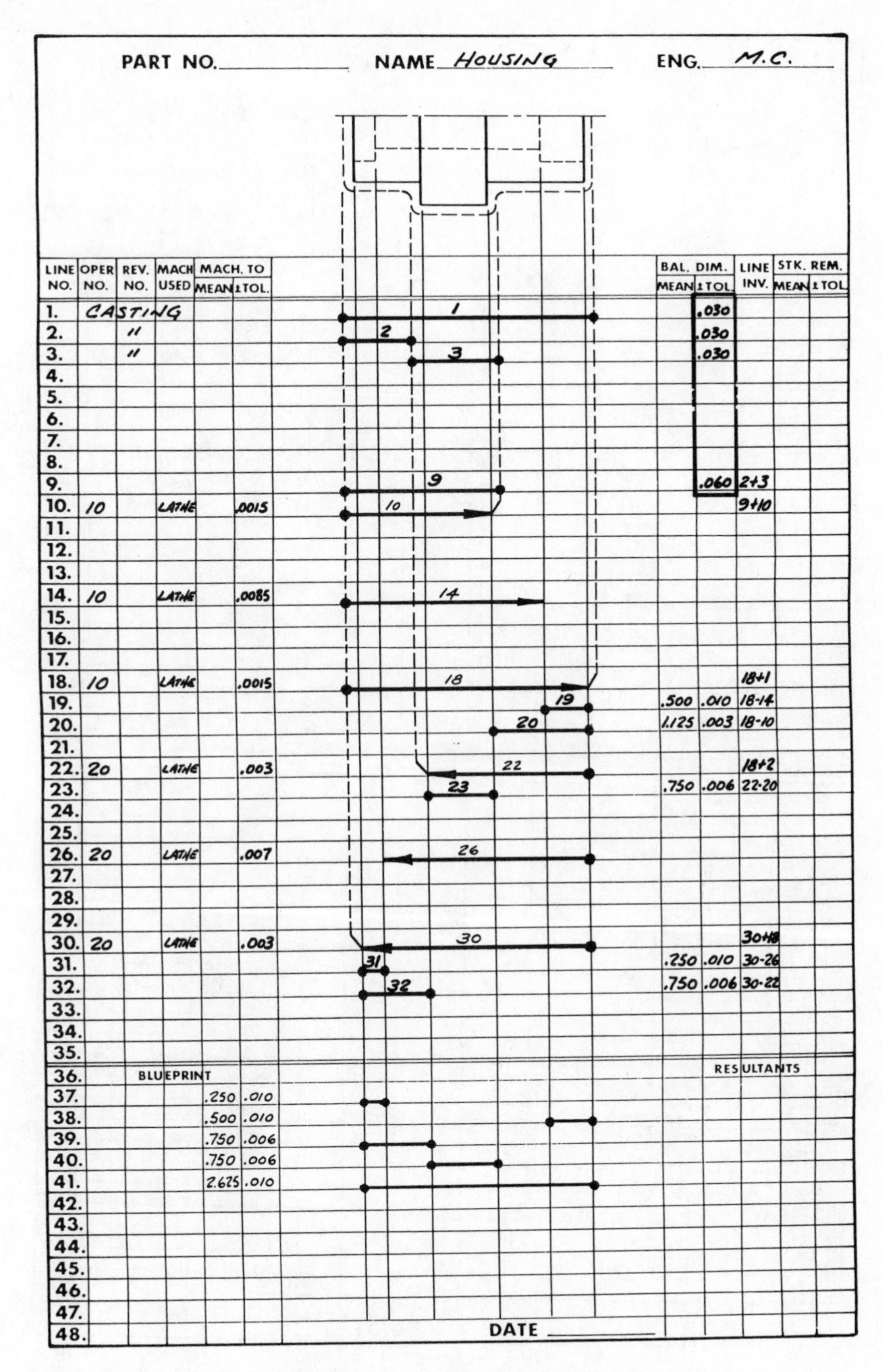

Figure 6-21 Dimension-and-Tolerance Computation, Step 5

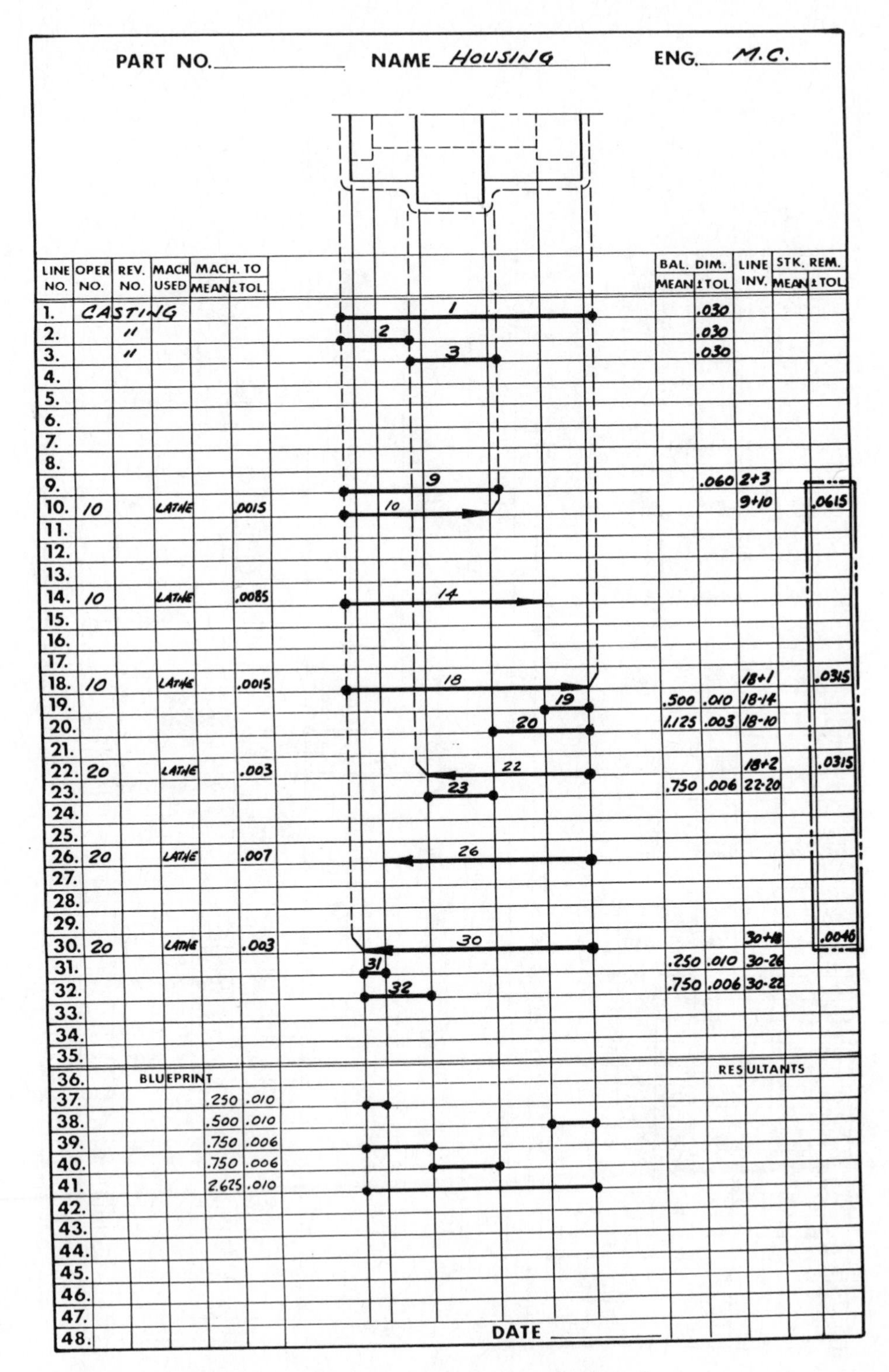

Figure 6-22 Dimension-and-Tolerance Computation, Step 6

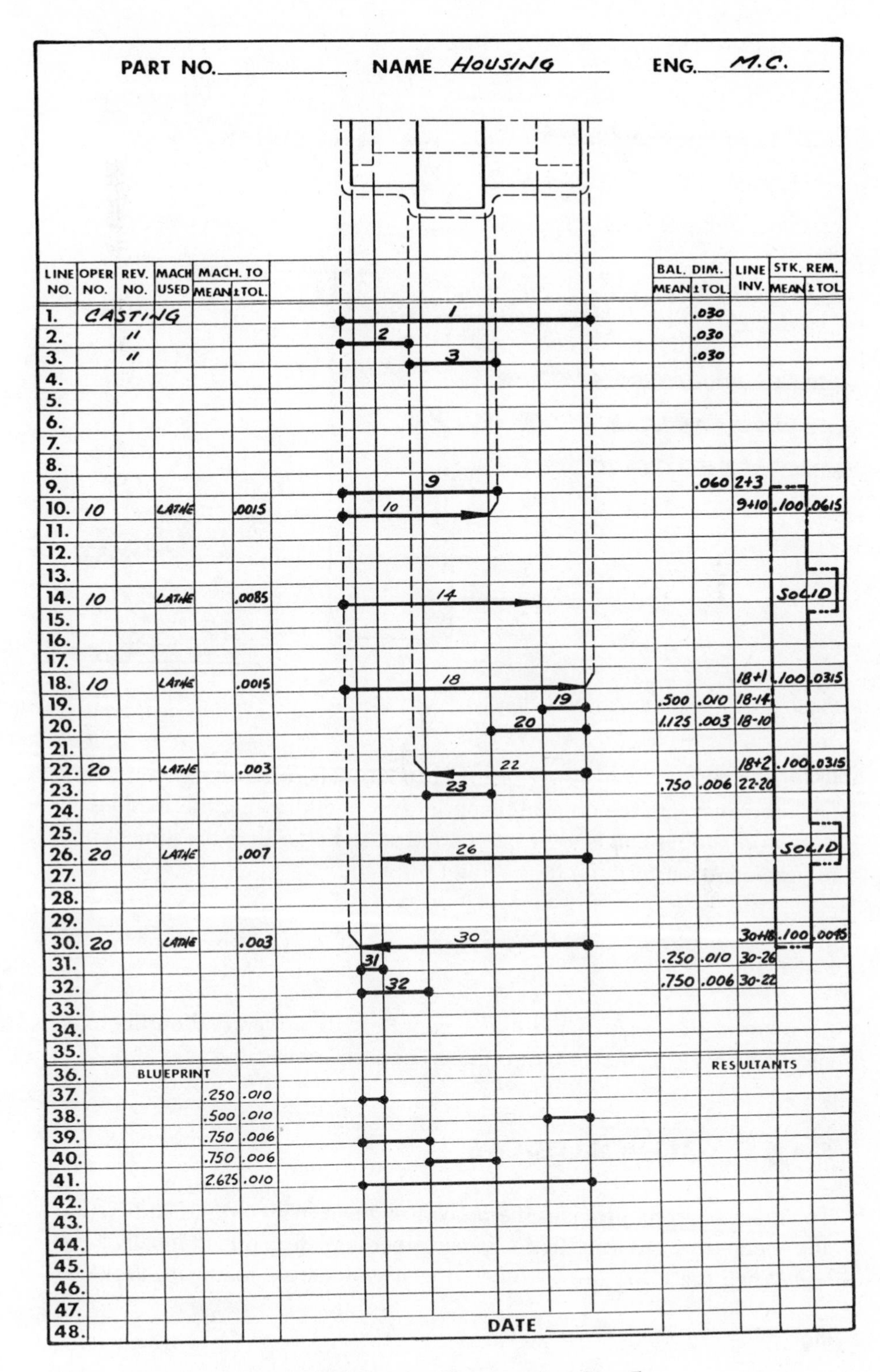

Figure 6-23 Dimension-and-Tolerance Computation, Step 7

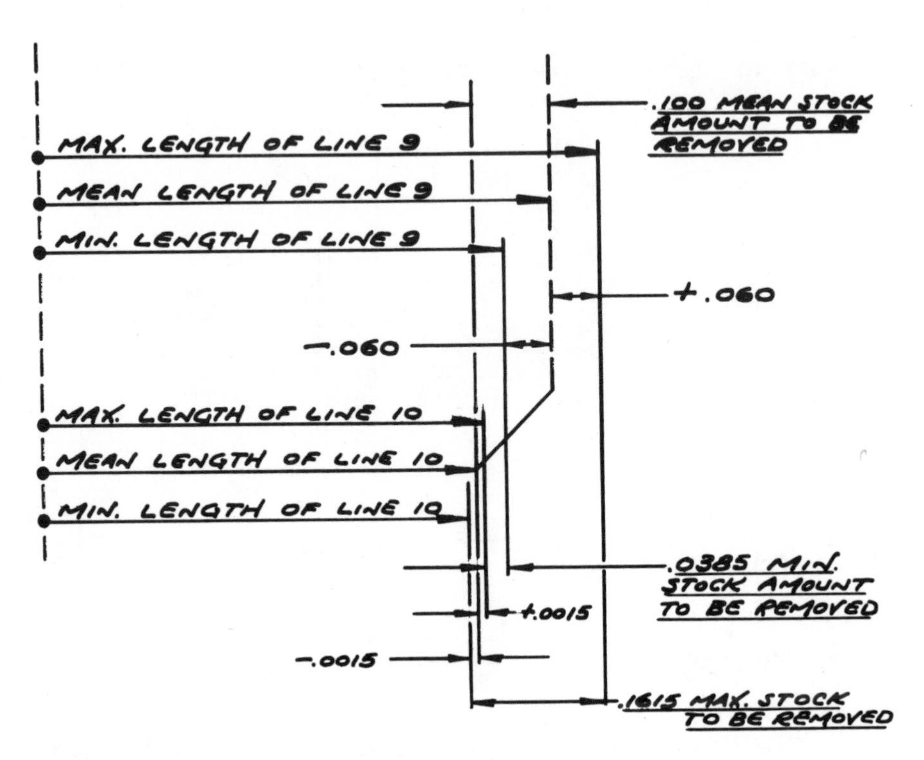

Figure 6-24 Stock Removal Rationale

finished blueprint dimension (see Fig. 6-26). These resultants must then be compared with the original blueprint (product drawing) dimensions. In each case, the resultants must be as good as or better than the dimensional requirements set forth on the product drawing.

STEP 10

All relevant dimensional and tolerance information recorded on the tolerance chart is now transferred to the original strip chart (see Fig. 6-27).

6 ▪ 5 CONCLUSION

Tolerance charting provides the process engineer with a rapid and systematic method of ensuring that a given process plan is dimensionally feasible. When the chart proves otherwise, it is a simple matter to reprocess

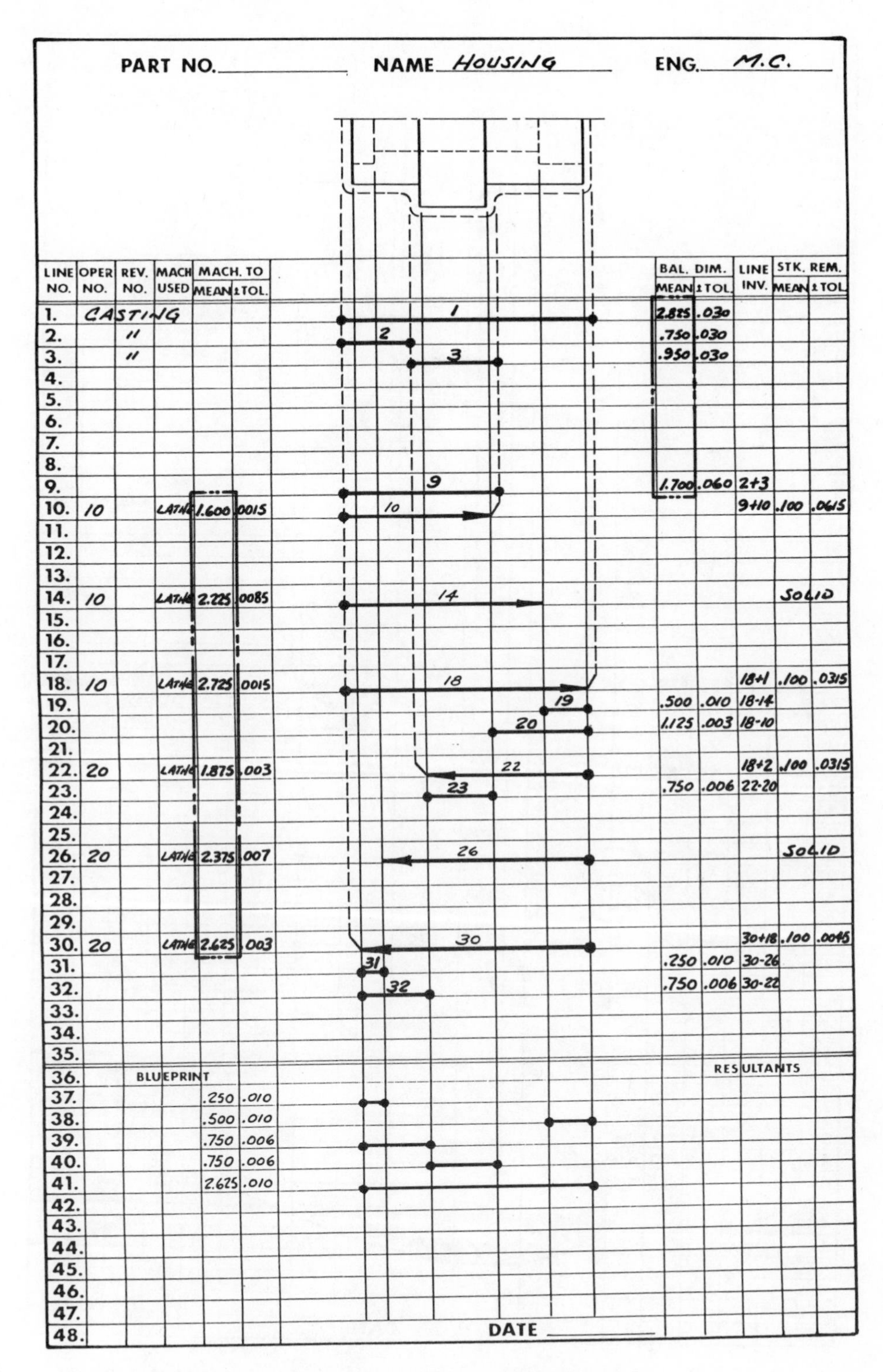

PART NO. ____________ NAME *HOUSING* ENG. *M.C.*

LINE NO.	OPER NO.	REV. NO.	MACH USED	MACH. TO MEAN	MACH. TO ±TOL.	BAL. DIM. MEAN	BAL. DIM. ±TOL.	LINE INV.	STK. REM. MEAN	STK. REM. ±TOL
1.	CASTING					2.825	.030			
2.		〃				.750	.030			
3.		〃				.950	.030			
4.										
5.										
6.										
7.										
8.										
9.						1.700	.060	2+3		
10.	10		LATHE	1.600	.0015			9+10	.100	.0615
11.										
12.										
13.										
14.	10		LATHE	2.225	.0085				SOLID	
15.										
16.										
17.										
18.	10		LATHE	2.725	.0015			18+1	.100	.0315
19.						.500	.010	18-14		
20.						1.125	.003	18-10		
21.										
22.	20		LATHE	1.875	.003			18+2	.100	.0315
23.						.750	.006	22-20		
24.										
25.										
26.	20		LATHE	2.375	.007				SOLID	
27.										
28.										
29.										
30.	20		LATHE	2.625	.003			30+18	.100	.0045
31.						.250	.010	30-26		
32.						.750	.006	30-22		
33.										
34.										
35.										
36.	BLUEPRINT					RESULTANTS				
37.				.250	.010					
38.				.500	.010					
39.				.750	.006					
40.				.750	.006					
41.				2.625	.010					
42.										
43.										
44.										
45.										
46.										
47.										
48.										

DATE ____________

Figure 6-25 Dimension-and-Tolerance Computation, Step 8

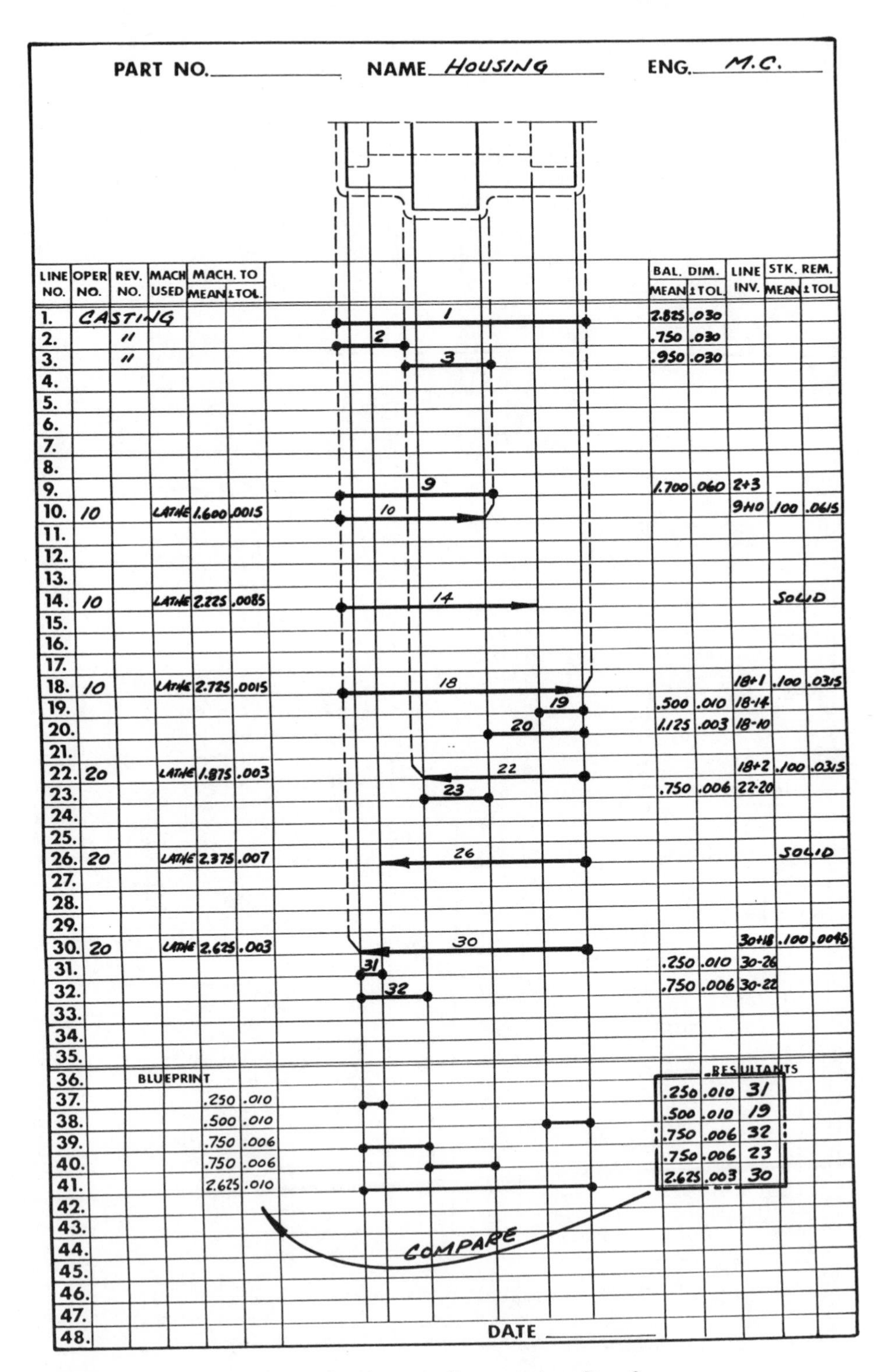

Figure 6-26 Dimension-and-Tolerance Computation, Step 9

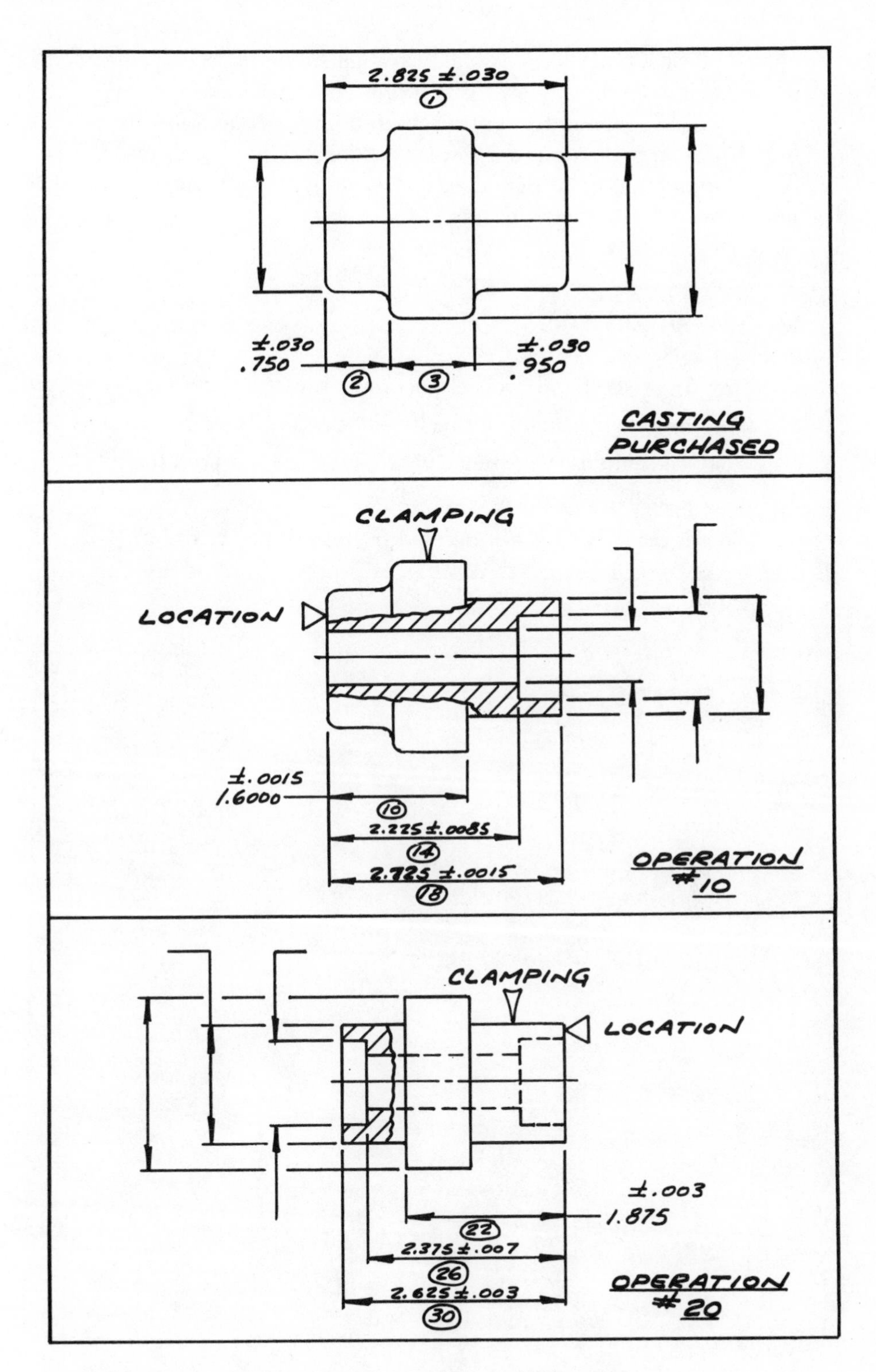

Figure 6-27 Strip Chart with Line Numbers and Dimensions

the part and check out its practicality by generating a second tolerance chart. In each case, the engineer is working with paper and not the usual scrap-generating "cut and try" approach used in many factories.

In today's competitive production environment, where quality and efficiency are more important than ever to the survival of any manufacturing concern, the use of tolerance charting is a must.

REVIEW QUESTIONS

1. Why do product and process tolerances sometimes differ?
2. What is tolerance charting specifically designed for?
3. What is a simplified tolerance chart and what is it used for?
4. What is a strip chart?
5. What is the major benefit derived from tolerance charting?

CHAPTER 7

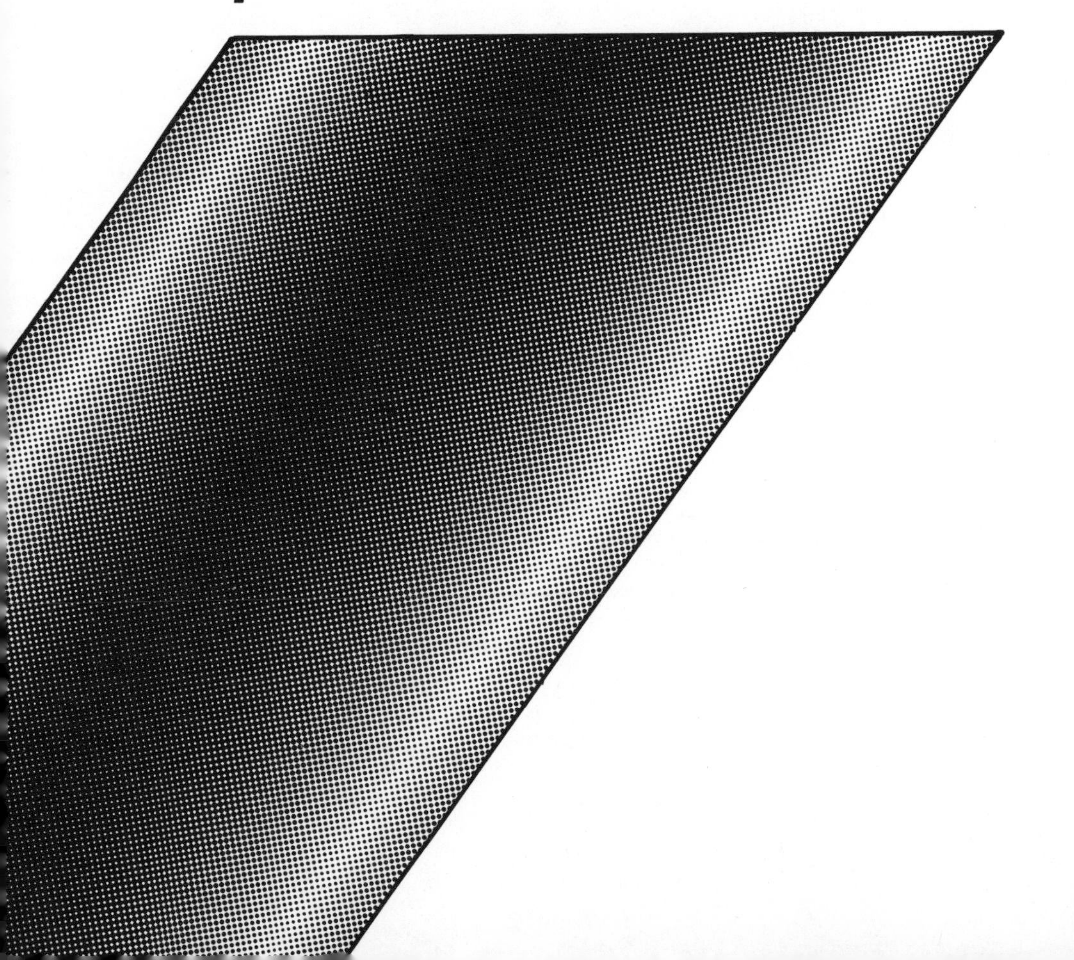

Computer-Aided Process Planning

7 ▪ 1 INTRODUCTION

Computer-Aided Process Planning (CAPP) is a relatively new technology that has been developed to increase the accuracy and productivity of the total manufacturing planning effort. CAPP is actually built upon three key elements: (1) existing, manual, process-planning techniques and knowledge as previously described in this text; (2) group technology (GT) theory, a manufacturing concept in which production parts are grouped according to their similarities; and (3) modern computer technology.

A variety of sophisticated CAPP software has been written to integrate these three elements with the following objectives in mind:

1. To improve the overall efficiency of process planning
2. To help planners develop accurate process plans and cost estimates
3. To capture the knowledge and experience of senior process planners
4. To eliminate or minimize the duplication of machinery, tooling, and engineering effort
5. To create a link with other computer data bases used in the company, such as CAD/CAM (Computer-Aided Design/Computer-Aided Manufacturing) and CNC (Computer Numerical Control)
6. To efficiently document all steps within the process-planning effort via computer-generated estimates, routings, and operation sheets.

CAPP is further broken down into two basic categories: that of the computer-assisted *variant* approach and the computer-assisted *generative* approach.

Briefly stated, the variant approach makes use of a historical data base of manually generated process plans. By retrieving and searching through these existing plans, the computer can assist the engineer in determining where variations on an old plan can be made to work for a new part.

The generative approach goes beyond simply storing possible answers to actually generating new process plans through the use of elaborate procedures, algorithms, if-then and or-else statements, and rules.

Both the variant and generative approaches to CAPP will be discussed at some length later in this chapter. However, before the subject of CAPP can be properly addressed, the planner must have a thorough understanding of group-technology theory as covered in the next section of this chapter.

Today's young process planners will undoubtedly be called upon to create, improve, and implement computer-aided process-planning systems. CAPP systems are often tailor-made to a given manufacturer's needs. Their complete implementation will be an evolutionary process, rather than an overnight one.

7 ▪ 2 GROUP TECHNOLOGY

Today, the term *group technology* is used to describe a manufacturing concept in which production parts are grouped according to their similarities.

These parts are typically grouped either by geometric characteristics or by their method of manufacture.

It is well known that per-part manufacturing costs go down when production volumes go up. By grouping small, odd, production lots together, manufacturing methods can be made to more closely approximate those of mass production. It is upon this premise that the entire philosophical foundation of group technology is built.

With the public's eye focused on mass-manufactured items such as cameras and automobiles, a major trend toward small-lot production has been obscured. This trend has recently been accelerated by efforts to reduce inventory levels through the use of just-in-time scheduling (JIT).

JIT is a philosophy fostered by original equipment manufacturers (OEMs) that ideally would permit less than one day's inventory of all needed production parts. This philosophy is forcing OEM suppliers to run smaller and smaller lot sizes and to ship only good parts on a daily basis.

Coupled with this trend toward JIT and smaller production lot sizes is the fact that a large percentage of the cost to produce a part rests in setup, inspection, material handling, and storage time. The reduction of these small-lot manufacturing costs lies in the ability to exploit part similarities through the use of group technology. Group technology, however, goes far beyond simply reducing production floor inefficiencies to sweeping implications for product design, process engineering, and tool design.

For product design, the need for new parts is often eliminated by the existence of a previously designed part that is already in production. The grouping and coding of all parts by their geometric characteristics permits the designer to search the system for those graphic parameters required in the new design. The result is typically fewer new designs and a saving of all the labor surrounding a new part in the system.

For process engineering, machine tools and processes can be selected with a group of parts in mind. The need for fewer unique routings and operation sheets is another benefit yielded through the grouping of parts.

The grouping of parts into families also serves to quantify the parameters of all the special tooling required. The tool designer can then strive to make new designs universal and capable of serving each member in the given group or family. Again, this reduces the number of new designs.

In short, group technology can be beneficial wherever large numbers of seemingly different items are produced in small lots. Before any of these benefits can be realized, however, parts must be classified, coded, and grouped.

Classification Systems

Systems of classification are developed for two specific yet interrelated reasons: (1) to enhance one's ability to retrieve a single classified item from the system and (2) to provide some useful information about each item in the system. An examination of three very common nonmanufacturing classification systems should serve to clarify this important concept.

The yellow pages of a phone book are classified by products and services. These broad classifications are then arranged alphabetically. If a person is interested, for example, in purchasing a new motorcycle, the key word *motorcycle* would lead them to a yellow pages listing of all of the motorcycle shops in the area. If one were to try to complete this same task using the white pages of the book it would be next to impossible.

A second example is that of our nation's postal zip code system. Numbers are used to identify regions of the country, states, cities, and even parts of cities. This numeric classification of the mail permits the rapid sorting and delivery of millions of individual letters and packages daily.

A third common classification system is found in our nation's many libraries in the form of the subject listing card catalog. Picture going into a library and trying to find a book on any subject without the books being classified in some manner; it would be an impossible task. By using the subject listing card catalog, all of that library's books on a given subject can be found in a matter of minutes. Of course this subject listing has a numeric classification system to back it up (i.e., the Dewey Decimal System or Library of Congress Classification System). In either case, the classification number assigned to the book contains specific information about the contents of the book and to which subject category or group it belongs.

The classification of manufactured parts is done for exactly the same reasons as in the nonmanufacturing examples outlined above. Again, classification of any item is done to enhance its retrieval from the system and to provide ready reference information about that item.

Manufactured parts are typically classified or grouped in one of four basic ways: (1) by part geometry, (2) by part function, (3) by manufacturing characteristics, and (4) by part material. These grouping techniques are covered in later sections. By taking and using these basic methods of classification, many proprietary computer-based systems have been developed. These systems are typically capable of accepting and classifying any number of highly diverse parts. However, the power and cost of such purchased systems will often go far beyond the needs and means of many smaller companies to whom group technology would still be beneficial. For these smaller companies, a small, specialized, and self-developed classification system can be the answer.

Fundamentally, all part classification systems are built around the assigning of code numbers to identify graphic and manufacturing characteristics of interest. Therefore, the assigning of code numbers to all of a plant's manufactured parts is the single most important step in the group technology implementation process.

Assigning Code Numbers

There are four recognized systems for the assignment of code numbers: (1) the monocode system, (2) the universal system, (3) the polycode system, and (4) the specific system.

Monocode systems are developed in a hierarchical manner: each digit assigned in the code number has an influence on the following digit to be assigned (see Fig. 7-1). Notice how the choice of the first digit in the code directs and restricts the choice of the second and so on.

An example code number yielded from the system shown in Figure 7-1 may be as follows: 1-1-2-2 + 053. This code number tells the reader that the part is round, with a diameter less than 5 in., with a length over 5 in., is made from cold-rolled steel, and is the 53rd part of this basic style designed.

Universal systems are those capable of accepting and coding a wide variety of part configurations. Some of the large computer-based universal systems could code and group any conceivable manufactured item. This versatility is often desired for use in extremely large and diverse companies and corporations.

Polycode systems are used to identify individual part characteristics: each digit assigned in the code number is independent of every other digit assigned (see Fig. 7-2). An example code number yielded from the system shown in Figure 7-2 might be as follows: 3-2-3-4-3-2-1-6 + 027. This code number tells the reader that the part has a 3.50-in. O.D., 2.25-in. B.C., .750-in.-dia. pilot hole, .437-in. flange thickness, 1.00-in. forging dia., .625-in. stem dia., .800-in. stem length, 2- to 2¼-in. shoulder length, and is the 27th shaft of this basic style designed.

The major advantage of the polycode system is the way that it can be charted and expanded without disturbing the location of existing coded variables.

GROUP NO. = □□□□ + SEQUENTIAL PART NO. (I.E. 001, 002, 003, ... 999)

PLACE WHERE NUMBER RESIDES

1		2		3		4	
NO.	DESCRIPTION	NO.	DESCRIPTION	NO.	DESCRIPTION	NO.	DESCRIPTION
1	ROUND (e.g. BAR STOCK)	1	DIA. < 5"	1	LENGTH < 5"	1	MATERIAL HRS
						2	" CRS
				2	" > 5"	1	" HRS
						2	" CRS
		2	DIA. > 5"	1	" < 10"	1	" HRS
						2	" CRS
				2	" > 10"	1	" HRS
						2	" CRS
2	RECTANGULAR (e.g. BLOCK)	1	THICK < 2" WIDE < 4"	1	" < 6"	1	" HRS
						2	" CRS
				2	" > 6"	1	" HRS
						2	" CRS
		2	THICK > 2" WIDE > 4"	1	" < 12"	1	" HRS
						2	" CRS
				2	" > 12"	1	" HRS
						2	" CRS
3	SHEET (e.g. SHEET METAL)	1	GAGE < 12	1	" < 60"	1	" HRS
						2	" CRS
				2	" > 60"	1	" HRS
						2	" CRS
		2	GAGE > 12	1	" < 80"	1	" HRS
						2	" CRS
				2	" > 80"	1	" HRS
						2	" CRS
4	OTHER (e.g. IRREGULAR SHAPED CASTING)	1	LBS < 10	1	REQUIRES MILLING	1	" STEEL
						2	" CAST IRON
				2	REQUIRES DRILLING	1	" STEEL
						2	" CAST IRON
		2	LBS > 10	1	REQUIRES MILLING	1	" STEEL
						2	" CAST IRON
				2	REQUIRES DRILLING	1	" STEEL
						2	" CAST IRON

Figure 7-1 A Universal Monocode Numbering System

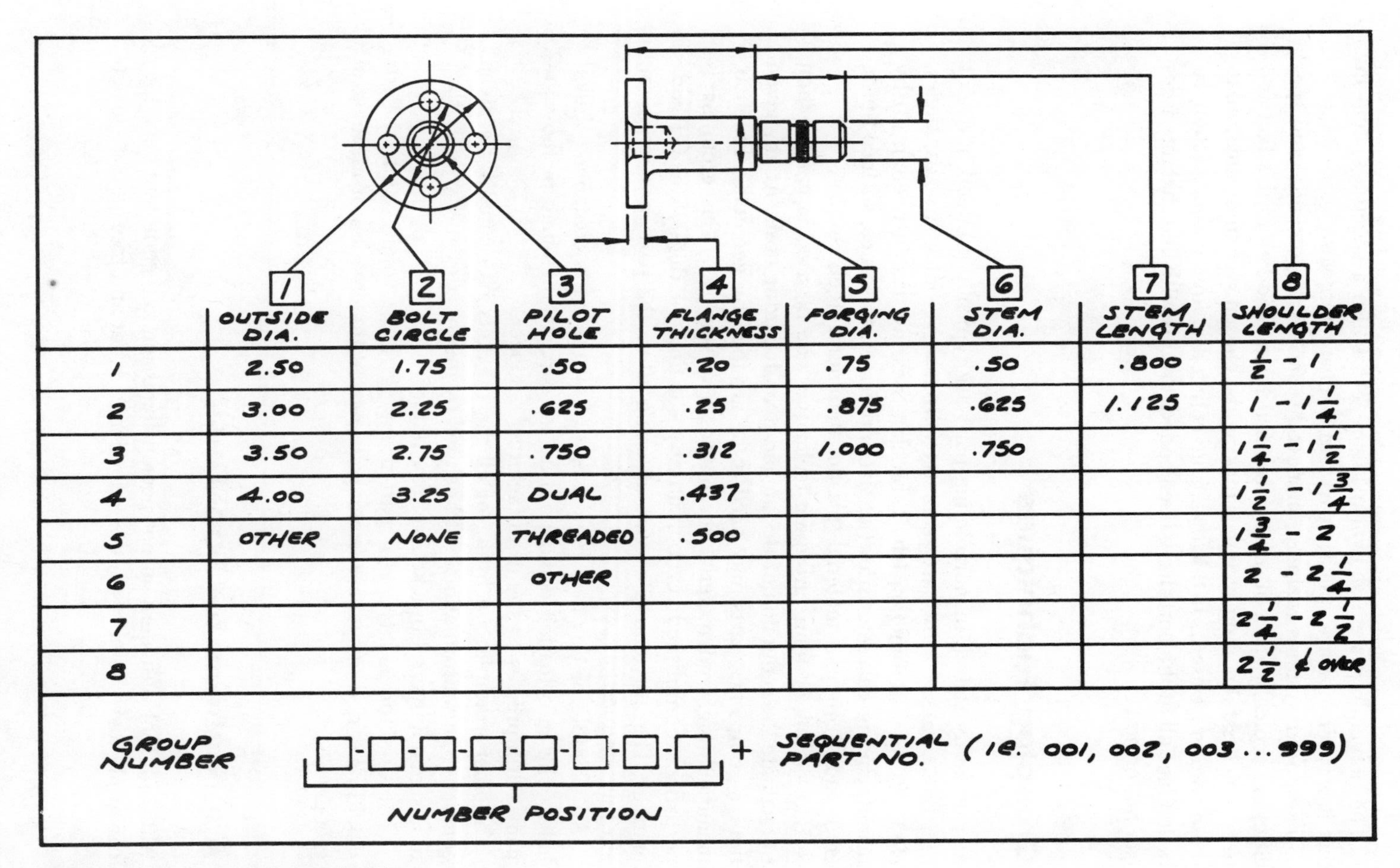

	1 OUTSIDE DIA.	2 BOLT CIRCLE	3 PILOT HOLE	4 FLANGE THICKNESS	5 FORGING DIA.	6 STEM DIA.	7 STEM LENGTH	8 SHOULDER LENGTH
1	2.50	1.75	.50	.20	.75	.50	.800	$\frac{1}{2}$ - 1
2	3.00	2.25	.625	.25	.875	.625	1.125	1 - 1$\frac{1}{4}$
3	3.50	2.75	.750	.312	1.000	.750		1$\frac{1}{4}$ - 1$\frac{1}{2}$
4	4.00	3.25	DUAL	.437				1$\frac{1}{2}$ - 1$\frac{3}{4}$
5	OTHER	NONE	THREADED	.500				1$\frac{3}{4}$ - 2
6			OTHER					2 - 2$\frac{1}{4}$
7								2$\frac{1}{4}$ - 2$\frac{1}{2}$
8								2$\frac{1}{2}$ & OVER

Figure 7-2 A Special Polycode System

The polycode system shown in Figure 7-2 is also a specific system in that it was set up for a specific part and its characteristics.

These four systems (namely, monocode, universal, polycode, and specific) are typically used in some combined form. The key to this fact and to any successful coding system is making the chosen system meaningful to those who must use it. If this means inventing a unique coding system, so be it. Once all of the parts have been coded, groups with like characteristics can be formed.

Grouping Techniques

As previously stated, manufacturing parts are typically grouped in one of four basic ways: (1) by part geometry (size, shape, holes, threads, angles, etc.); (2) by part function (shaft, bracket, bearing, fastener, etc.); (3) by manufacturing characteristics (lot size, processes, tolerance, surface finish, etc.); and (4) by part material (plastic, aluminum, steel, etc.).

Like coding systems, grouping techniques can be used in combination to facilitate the reduction of design and manufacturing costs. At any given time, a company's parts may need to be grouped by geometry, function, manufacturing, and material for purposes of analysis. This grouping is normally accomplished via computer by isolating one digit of the GT code number. The outcome is a printout listing all of the parts possessing the attribute in question.

An example of this may be a listing of all the parts containing drilled holes. This information could be used for capacity planning of the company's drill presses by the manufacturing engineer.

Another example is a listing of all the brackets fabricated by a given company. This information could be used by the tool designer who is asked to design a universal bracket-welding fixture.

The uses for and types of part groupings are virtually endless. Parts should be coded in such a manner as to permit the creation of meaningful groups.

Applications for Tool Design

Although tool designers do not personally assign GT code numbers, they can apply these numbers in the reduction of labor and tooling costs generated by their department. Specifically, these dollar savings come as a result of the following GT applications:

1. Group technology provides parameters for the design of universal tooling.
   A. Less setup labor
   B. More machine uptime
2. Group technology enhances the search for usable existing tooling.
3. Fewer new or individual tooling designs are required with group technology.
   A. Less tool design labor
   B. Fewer tooling fabrication dollars
4. Fewer design mistakes of omission (such as, forgetting one or more parts that the tooling must accommodate).
   A. Less redesign labor
   B. Less repair labor
5. Fewer tool layouts are required.

The benefits of group technology are so far-reaching, both within tool design and without, that it would seem logical for all companies to have embraced the concept. This, however, is not the case. The majority of manufacturing companies in the United States have not even heard of, or do not understand the concept of group technology. If a tool designer or process engineer finds himself or herself working for a company that has yet to discover the benefits of group technology, he or she should take the initiative to propose and explain the concept. (Curtis, *Tool Design,* 1986)

7 ▪ 3 CAPP

Variant Process Planning

Variant process planning is a natural extension of manual process planning in which the computer is employed. With the variant system, all existing process plans are classified, coded, and stored in the computer's memory. These classifications and codes are usually based upon group-technology theory in that they relate to specific geometric characteristics found in families of parts. Existing operational sequences and individual process parameters are also coded for computer retrieval. With such information, variant system software can be used to examine a new part in terms of existing routings and manufacturing capabilities.

The output of any variant system is a list of those existing process plans (routings) that most closely match the processing requirements of the new part. Here, manual process planning reenters the picture, as the engineer, with his or her processing experience, is required to analyze the existing plans that were retrieved by the computer. In some cases, an existing plan can be used for a new part. In other cases, an old plan must be modified by adding or rearranging operations. If no similar plan exists, the process engineer must create a new plan from scratch.

As each new or modified plan is generated, it too is coded and entered into the system for future access.

Generative Process Planning

The generative process-planning approach differs from the variant approach in that no standard or existing plans are predetermined and stored in the computer's memory. Rather, through the use of some very sophisticated software, the computer automatically creates new routings and unique, individual, process sheets for each part estimated or released for production by manufacturing.

Generative process planning requires two essential elements. The first element is a coding system that translates the physical geometry of the part, as shown on the product drawing, into data that can be interpreted by the computer. The part's rough and finished state—including all datums, dimensions, and tolerances—must be coded in terms that relate to existing in-house machine capabilities. Information about existing durable and perishable tooling is also coded and entered into the system. This coded generative machine and tool data must be continually updated to reflect the ongoing changes (such as, additions and deletions) taking place on the shop floor. The coding requirements for a generative process-planning system are far more detailed than those required in a variant system.

The second element required in a generative process-planning system is the software. This software uses logic, formulas, algorithms, and rules to check the part's geometric or processing requirements against existing in-house machine-and-tooling capabilities. The software actually selects all operations, specific machine tools, and tooling required in the processing of a new part. Cycle times, production rates, and setup hours are also calculated for each operation. In addition, when interfaced with CAD/CAM, generative process-planning software can generate formal process sheets, tool layouts, and numerical control (NC) tool-path information.

Design, development, and implementation of a generative process-planning system is a multimillion dollar proposition involving literally many man-years of engineering labor. Because of this fact, generative process planning is today more theory than practice, with a precious few large corporations having the financial wherewithal to start such a venture. As of the mid 1980s, no more than a dozen companies and academic institutions had engaged in generative process-planning systems development. However, a 1980 Delphi forecast, commissioned by and through the Society of Manufacturing Engineers, predicted that by 1990 50 percent of all process plans made in the United States will be computer generated.

7 ■ 4 CONCLUSION

The future of group technology and CAPP can be pondered through an examination of contemporary industrial events. Although related, these events have appeared as independent reactions to individual problems. An outline of these events is listed below.

1. Strong foreign competition in the late 1970s and early 1980s forced American industry to search for ways to increase productivity and quality.
2. Correct process selection is receiving emphasis in an effort to increase quality and reduce scrap.
3. Just-in-time scheduling has become a popular experiment in inventory control.
4. Many companies are implementing the theory of group technology with more than modest success.
5. Increasingly sophisticated computer software is being developed with the ability to tie together GT, CNC machine tools, process planning, CAD, and inventory control.
6. Computer-aided process planning (CAPP), the selection of part-processing methods via computer, has become a reality.
7. Efforts toward the creation of the totally automated factory have been made in an effort to increase productivity and quality.

Notice how each of the events outlined above centers around the question of increasing productivity and quality. Treated as independent variables, these events do not possess the strength necessary to have any more than a minor impact on the problem. The future of process planning, if not the whole future of the American manufacturing industry, rests in a single, integrated computer-based system that totally embraces the concepts of group technology, just-in-time scheduling, CNC machining, CAD, and CAPP. (Curtis, *Tool Design*, 1986)

REVIEW QUESTIONS

1. What are the three key elements upon which computer-aided process planning is built?
2. What are the major objectives of computer-aided process planning?
3. How do the variant and generative approaches to process planning differ?
4. What is group technology?
5. What are some of the major grouping techniques used in the implementation of group technology?

CHAPTER 8

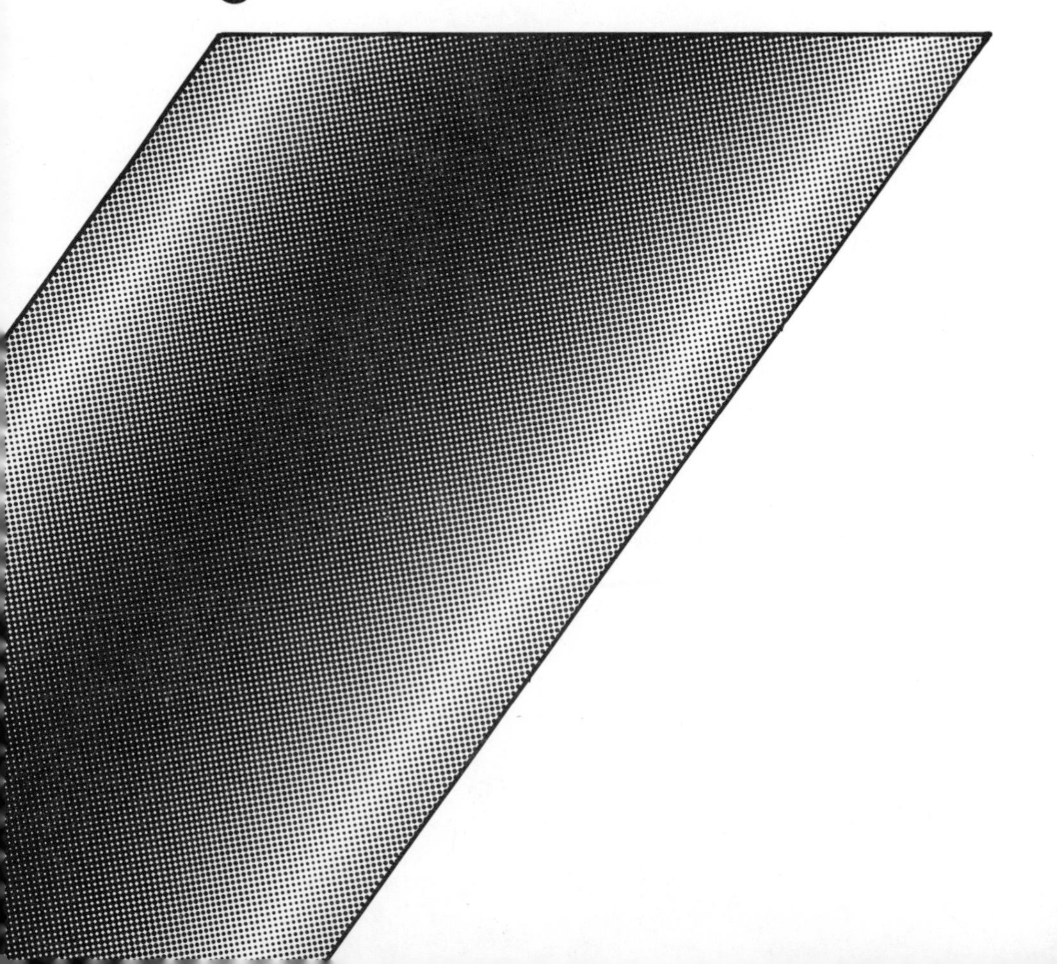

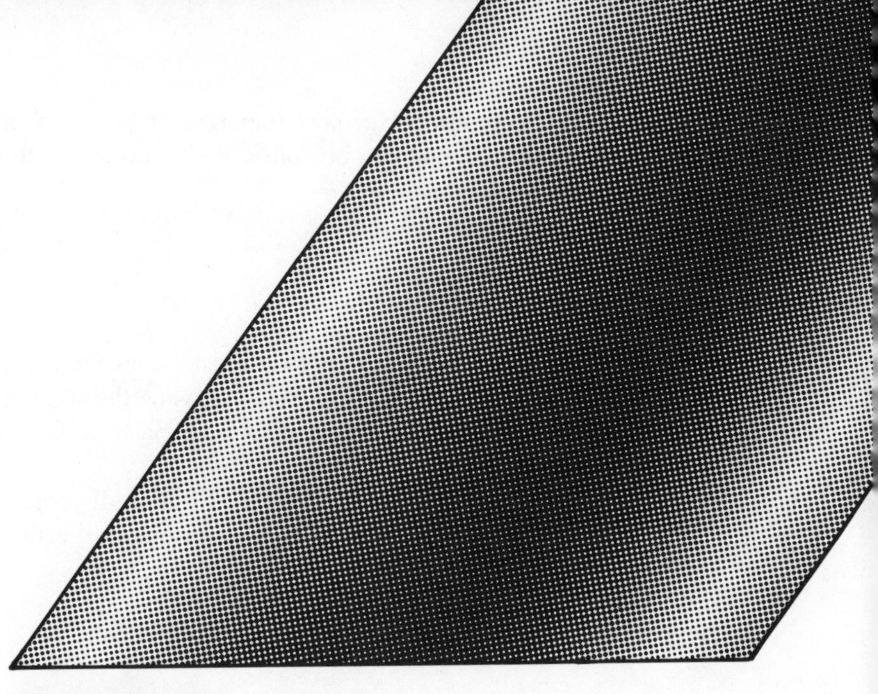

Project Management

8 ▪ 1 INTRODUCTION

In the first seven chapters, process planning has been explained and treated as an independent task. Independent, in the sense that once the process plan is complete the engineer's job is over. However, process planning is an integral part of each event required to take a new product from the release stage to the production stage. Because it is so much a part of every manufacturing detail and decision many companies, from the outset, treat each new product release as a project. In such instances, the process planner often takes on the new title of project manager or project engineer. No discussion of process planning can be considered complete without a brief discussion of project management.

8 ▪ 2 THE PROJECT

The project or planned undertaking that must be managed, supervised and coordinated is made up of many steps between the signing of a new product sales agreement and the shipping of good parts to the customer by a promised deadline.

A review of the steps typically involved in this kind of project is now in order. These steps are listed in approximate chronological order.

PROJECT STEPS

1. A formal sales agreement between the customer and the vendor is signed.
2. A formal new product release document is sent to all affected departments authorizing the use of appropriate resources to meet the terms of the sales agreement.
3. The process engineer pulls and reviews the cost estimate transmittal, that is, the preliminary process plan.
4. Make or buy decisions are firmed up and prints of buy parts are sent to purchasing.
5. A tentative routing is made for each make part including machine selection.
6. A strip chart is made for each make part as is appropriate.
7. A formal process sheet is made for every operation on each part and includes the specification of all tooling and gauging required.
8. Design work orders and purchase requisitions are written to cover required machinery and tooling.
9. Design work and purchase orders are calculated for management approval. This may and should also include a request for a formal plant layout.
10. Orders are sent to an in-house tool design department or out for tool and machine design.

11. Tool designs are checked and approved.
12. Tool designs are sent to the toolroom or to an outside tool shop for build.
13. Plant layout is approved.
14. Machines and tools are tried out and approved for shipping to the plant.
15. Tools/machines are received and inspected at the plant.
16. Tools/machines are moved to the appropriate plant location according to the plant layout or are put into storage.
17. Tools/machines are set up, tried out and debugged.
18. Inspection samples are run off at each operation and inspected to determine the capability of the setup.
19. If the setup proves to be capable, it is released for production; if not, it is back to Step 17.
20. Production samples are run, numbered, and sent to the customer for approval.
21. Production may start upon customer approval of the samples.
22. Industrial engineering is then brought in to check and set production standards that will replace the previously estimated standards.
23. All changes in processing, tooling, and so forth, must now be documented for future reference.

8 ▪ 3 MANAGEMENT

The project steps are the means by which the shipping of good parts may be accomplished on time. Management, in this case, is simply the judicious use of these steps.

The Critical Path

Once a project is broken down into a number of steps or activities, the time required to complete each event can be estimated. With these individual estimates, a minimum time can be determined to complete the whole project.

In their proper sequence, the project steps create a path of connected

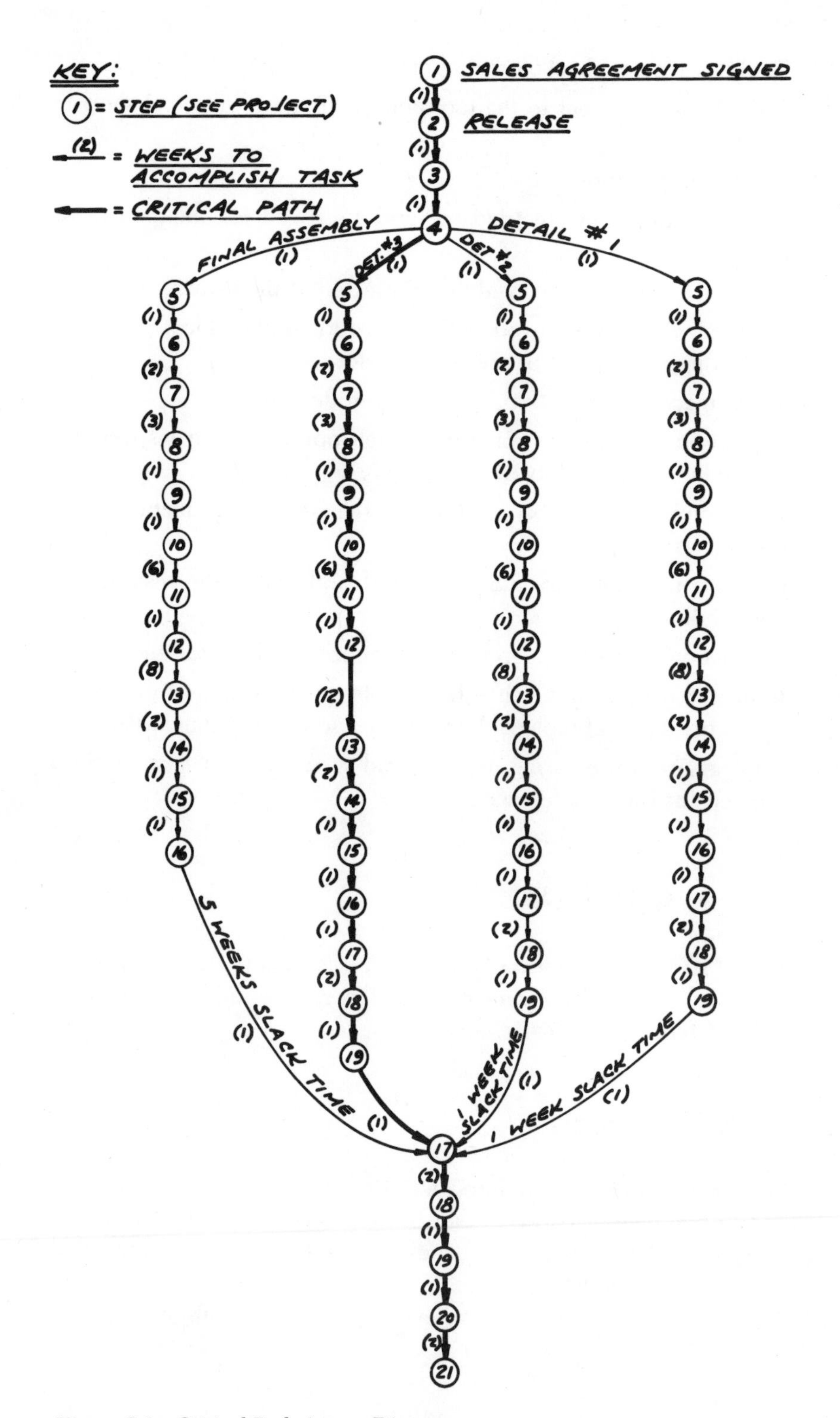

Figure 8-1 Critical-Path Arrow Diagram

activities. There may be many such paths in any project activity network; however, it is the length of the longest path that predicts the duration of the project and which is referred to as the *critical path*.

To illustrate the concept of the critical path, an arrow diagram can be drawn (see Fig. 8-1). In this example, three machined details come together in one final assembly which is the product to be sold. The production Detail 3 demonstrates the completion of Steps 17 to 21 of the project steps for the final assembly, thereby placing the critical path along Detail 3.

A project manager can monitor progress along the critical path in an effort to manage and control the outcome. If the length of the critical path begins to grow, the project manager can use additional resources (such as money and personnel) to bring it back in line. If sufficient slack time exists at the end of the critical path, the path itself may be allowed to grow within the limits of the extra or slack time.

8 ▪ 4 CONCLUSION

With any major project there are uncertainties; however, it is the role of the project manager to identify potential problems before they threaten to jeopardize the delivery of quality products on time.

If any manufacturing company is to meet new production obligations and thereby stay in business, each implementation plan must be reduced to a network of activities. The duration of each activity must be accurately estimated and monitored by a project evaluation system.

REVIEW QUESTIONS

1. ✓ Why are process planners sometimes called project managers or project engineers?
2. ✓ What is the project that must be managed by the process planner and why?
3. What is a critical path?
4. How can a critical path be shortened?
5. Is it always necessary to worry about the critical path getting longer? Why or why not?

APPENDIX I

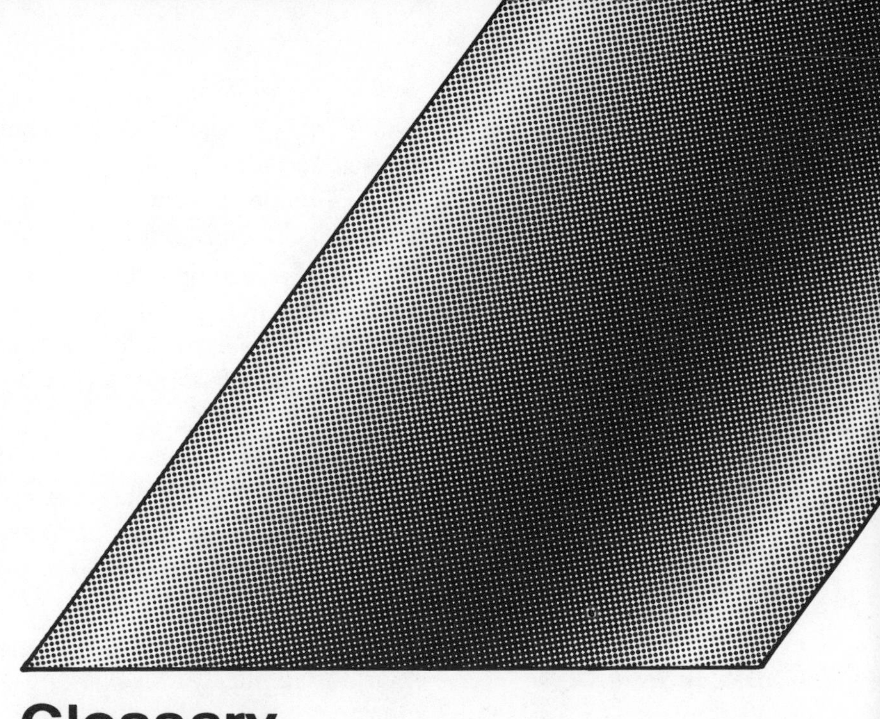

Glossary

Appropriation Type Work Orders — Means that company funds are set apart or assigned for a particular purpose or use and relate to work orders written to accomplish this purpose.

Capital Equipment — One or more machine tools (e.g., a drill press, lathe, milling machine, grinder, etc.) designed to be capable of accommodating a certain type of work over a wide range of sizes.

CAPP — Computer-Aided Process Planning.

CET — Cost Estimate Transmittal; a document that provides the process planner with the pertinent information related to the required accuracy and urgency of the estimate.

Design Work Order — A written request for design and or toolroom labor, including materials and the necessary funds required to cover the same.

Direct Labor — The actual cost of labor incurred in the processing of parts. Direct labor adds value to the product.

Durable Tooling — The portion of the setup (hardware) that will last for years under normal operating conditions in industry (i.e., lathe chucks, tool holders, jigs, fixtures, gauges, etc.).

ECN — Engineering Change Notice; a formal document that authorizes a change to be made on an existing product drawing. It also has implications for all related tooling.

ECR — Engineering Change Request; a document that precedes the ECN and requests approval of proposed product drawing changes from all affected departments.

Existing Machinery — Refers to those pieces of capital equipment presently owned by the process planner's company.

Expensed Type Work Orders — Work orders written to maintain normal operations within the plant (i.e., repair or replacement of worn out equipment).

Fixed Overhead — This is made up of costs that tend to remain independent from increases or decreases in total plant production.

Group Technology — A term used to describe a manufacturing concept in which production parts are grouped according to their similarities.

Indirect Labor — All labor that is not direct labor (i.e., material handling). Indirect labor does not add to the value of the product.

In-Process Operations — Operations that happen automatically, often en route between other operations; they involve little or no direct labor.

Machine Utilization — Expressed as a percentage, this is the portion of straight-time hours per month that a machine is presently being used.

Perishable Tooling — The portion of the setup (hardware) that will wear out and require replacement over and over again (e.g., drill bits, end mills, grinding wheels, etc.).

Process Planner — A manufacturing and process-engineering specialist who is responsible for the conception, planning, and implementation of economically justifiable production processes designed to produce a variety of industrial and consumer goods.

Process Planning — The devising of a particular method of manufacturing, generally involving a number of steps or operations.

Process Sheets — Also called operation sheets, these are complete sets of instructions utilized by setup, production, and inspection personnel in the normal course of their work.

Purchase Order — A legal document authorizing goods or services to be delivered and paid for.

This follows a purchase requisition (see below).

Purchase Requisition A formal written request for something to be purchased. This precedes a purchase order (see above).

Release A formal document that authorizes the expenditure of company resources on a given job.

Routing A prearranged list of the order in which operations (a process plan) are to be executed. It is normally the first sheet of the process drawings.

Tolerance Chart A document that provides the engineer with a systematic way in which length dimensions and tolerances can be determined and maximized at each operation. It is also a record of how such dimensions and tolerances were determined.

Tool Drawing In various forms, these are semipermanent records of tooling and include assembly and detail drawings.

Tooling All hardware employed in the manufacture of consumer and industrial goods.

Tool Layout A drawing that shows the arrangement of all tooling used in a specific operation or setup.

Tool Order See Design Work Order.

Variable Overhead This is made up of the costs that tend to identify or vary with total plant production.

APPENDIX II

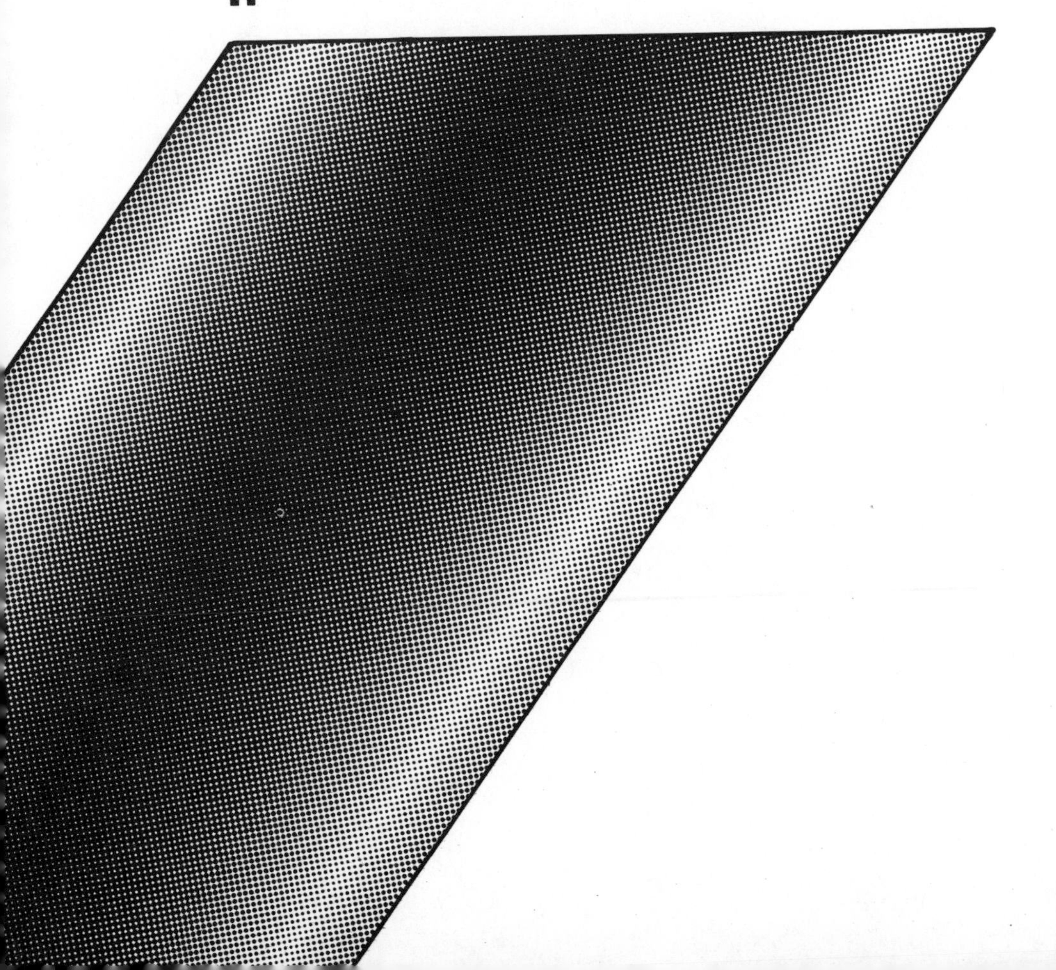

Materials

Table 1 Carbide Cutting Tools

MATERIAL	APPLICATION INFORMATION		
CARBIDE CUTTING TOOLS TYPES	1. PRIMARY USE	2. PRIMARY ATTRIBUTE	3. MATERIAL BEING MACHINED
1. C-1	ROUGHING	MEDIUM HIGH SHOCK AND MEDIUM LOW WEAR RESISTANCE	CAST IRON. NONFERROUS AND NONMETALLIC MATERIALS
2. C-2	GENERAL PURPOSE	MEDIUM SHOCK AND MEDIUM WEAR RESISTANCE	CAST IRON. NONFERROUS AND NONMETALLIC MATERIALS
3. C-3	FINISHING	MEDIUM LOW SHOCK AND HIGH WEAR RESISTANCE	CAST IRON. NONFERROUS AND NONMETALLIC MATERIALS
4. C-4	PRECISION FINISHING	LOW SHOCK AND HIGH WEAR RESISTANCE	CAST IRON. NONFERROUS AND NONMETALLIC MATERIALS
5. C-5	ROUGHING	EXCELLENT RESISTANCE TO CUTTING TEMPERATURE AND SHOCK LOADING	STEEL AND STEEL ALLOYS
6. C-6	GENERAL PURPOSE	MEDIUM HIGH SHOCK AND MEDIUM LOW WEAR RESISTANCE	STEEL AND STEEL ALLOYS
7. C-7	FINISHING	MEDIUM SHOCK AND MEDIUM WEAR RESISTANCE	STEEL AND STEEL ALLOYS
8. C-8	PRECISION FINISHING	LOW SHOCK AND VERY HIGH WEAR RESISTANCE	STEEL AND STEEL ALLOYS
9.			
10.			
11.			
12.			

APPLICATION INFORMATION			
4. ISO CLASS	5. ISO COLOR MARKING	6. MAJOR MANUFACTURERS	7. SPECIFIC OPERATIONS AND WORKING CONDITIONS
K30, K40	RED	CARBOLOY, SANVIK, KENNEMETAL, TRW, VALENITE	TURNING AND MILLING UNDER UNFAVORABLE CONDITIONS
K20	RED	CARBOLOY, SANVIK, KENNEMETAL, TRW, VALENITE	TURNING, MILLING, BORING, REAMING, COUNTERSINKING, AND BROACHING
K10	RED	CARBOLOY, SANVIK, KENNEMETAL, TRW, VALENITE	TURNING, MILLING, BORING, REAMING, COUNTERSINKING, AND BROACHING
K01	RED	CARBOLOY, SANVIK, KENNEMETAL, TRW, VALENITE	PRECISION TURNING, BORING, AND MILLING
P40, P50 M40	BLUE YELLOW	CARBOLOY, SANVIK, KENNEMETAL, TRW, VALENITE	TURNING, MILLING, ETC. UNDER UNFAVORABLE CONDITIONS
P20, P30 M20, M30	BLUE YELLOW	CARBOLOY, SANVIK, KENNEMETAL, TRW, VALENITE	TURNING, MILLING, ETC.
P10 M10	BLUE YELLOW	CARBOLOY, SANVIK, KENNEMETAL, TRW, VALENITE	TURNING, MILLING, ETC.
P01	BLUE	CARBOLOY, SANVIK, KENNEMETAL, TRW, VALENITE	HIGH PRECISION TURNING AND BORING

Table 2 Cutting Fluids

MATERIAL	APPLICATION INFORMATION		
CUTTING FLUIDS	1. PRIMARY USE	2. PRIMARY ATTRIBUTE	3. CHEMICAL ADDITIVES
TYPES			
1. EMULSIFYING OR SOLUBLE OILS	MACHINING OPERATIONS	LUBRICATION AND COOLING	OIL
2. REGULAR SOLUBLE OIL	MACHINING OPERATIONS	COOLING	OIL
3. PASTE COMPOUNDS	GRINDING	LUBRICATION AND COOLING	SAPONIFIED MINERAL OIL
4. STRAIGHT MINERAL OIL	LIGHT MACHINING OPERATIONS ON AUTOMATICS	COOLING, LUBRICATION, AND PENETRATING	MINERAL OIL
5. STRAIGHT FATTY OIL	HEAVY MACHINING OPERATIONS	CHIP LUBRICATION	ANIMAL FAT
6. MINERAL LARD OIL	HEAVY MACHINING OPERATIONS	LUBRICATION AND COOLING	MINERAL OIL AND LARD OIL
7. SULFURIZED AND CHLORINATED CUTTING OIL	MACHINING ALLOY STEEL	PRODUCES METALLIC FILM LUBRICATION	SULFURIZED OIL, CHLORINATED OIL, AND MINERAL OIL
8. CHEMICAL OR SYNTHETIC	TAPPING, THREADING, SAWING, BROACHING, AND GEAR CUTTING	EXCELLENT LUBRICITY AND RUST PREVENTION	VARIOUS
9. AQUEOUS SOLUTIONS	TURNING, MILLING, ETC.	COOLING AND CHIP FLUSH	CARBONATE OF SODA, BORAX, CAUSTIC SODA, LARD OIL, AND SOFT SOAP
10.			
11.			
12.			

APPLICATION INFORMATION			
4. APPLICATION METHOD	5. MIXTURE	6. RELATIVE COST	7. DISADVANTAGE
MANUAL, FLOOD, OR MIST	1 PART OIL 5 PARTS WATER TO 1 PART OIL 100 PARTS WATER	LOW	PARTS MAY RUST
MANUAL, FLOOD, OR MIST	1 PART OIL 5 PARTS WATER TO 1 PART OIL 100 PARTS WATER	LOW	MACHINE COOLANT SYSTEM MUST BE CLEANED PRIOR TO USE
MANUAL, FLOOD, OR MIST	———	LOW	HOT WATER MAY BE REQUIRED FOR MIXING
MANUAL, FLOOD, OR MIST	VARIES	MEDIUM	NONE
MANUAL, FLOOD, OR MIST	———	HIGH	TENDS TO BECOME RANCID, HIGH BACTERIA COUNT
MANUAL, FLOOD, OR MIST	10 - 40% MINERAL OIL	LOW	NONE
MANUAL, FLOOD, OR MIST	———	HIGH	OPAQUE OIL PREVENTS PART VIEWING
MANUAL, FLOOD, OR MIST	PROPRIETARY	MEDIUM - HIGH	TENDS TO FOAM
MANUAL, FLOOD, OR MIST	VARIES	LOW	OBJECTIONABLE SMELL, HOWEVER MAY BE CURED BY ADDING UNSLAKED LIME

Table 3 Iron

MATERIAL	APPLICATION INFORMATION		
IRON TYPES	1. PRIMARY USE	2. PRIMARY ATTRIBUTE	3. CHEMICAL COMPOSITION
1. CAST	THIS IS A GENERIC CLASSIFICATION	SEE SPECIFIC TYPE OF IRON (I.E. GRAY)	C > 2.00%
2. COMPACTED GRAPHITE	FLYWHEELS, BEARING CAPS, AND HYDRAULIC COMPONENTS	MACHINABILITY AND THERMAL CONDUCTIVITY	HAS A STUBBY INTERCONNECTED GRAPHITE FLAKE STRUCTURE
3. GRAY CAST	ENGINE BLOCKS, CYLINDER HEADS, MANIFOLDS, ETC.	VARIES WITH THE SIZE OF THE GRAPHITE FLAKES	GRAPHITE FLAKES ARE PRESENT IN THE IRON
4. HIGH ALLOY	APPLICATIONS THAT TAKE ADVANTAGE OF THE PRIMARY ATTRIBUTES	CORROSION, HEAT AND WEAR RESISTANCE	3 -40% ALLOYING ELEMENT
5. INGOT	CULVERTS, FLUMES, ROOFING, AND SIDING	PAINTABLE, WELDABLE	PURE IRON
6. MALLEABLE	HAND TOOLS, UNIVERSAL YOKES, AND AXLE HOUSINGS	STRONG, DUCTILE, FATIGUE STRENGTH AND MACHINABILITY	FERRITIC OR PEARLITIC
7. NODULAR (DUCTILE)	CRANKSHAFTS, CONNECTING RODS, GEARS, STEERING KNUCKLES, ETC.	STRENGTH, DURABILITY, TOUGHNESS, AND GOOD MACHINABILITY	GRAPHITE NUCLEATES INTO SPHERES
8. PIG	USED IN MAKING STEEL	REFINES TO ALSO PRODUCE WROUGHT AND INGOT IRON	Fe
9. WHITE CAST	APPLICATIONS THAT TAKE ADVANTAGE OF THE PRIMARY ATTRIBUTES	HIGH COMPRESSIVE STRENGTH AND EXCELLENT WEAR RESISTANCE	CEMENTITE DOMINATES THE MICROSTRUCTURE
10. WROUGHT	PIPE	FORMABLE, WELDABLE	CONTAINS A CONSIDERABLE AMOUNT OF SLAG
11.			
12.			

APPLICATION INFORMATION			
4. STRENGTH T - TENSILE S - SHEAR Y - YIELD (THOUSANDS OF PSI)	5. WEIGHT PER CUBIC FOOT (LBS.)	6. RELATIVE COST	7. DISADVANTAGE
T -- 20 - 60 S -- 1 - 1.6T Y -- N/A	438.7 - 482.4	MEDIUM	NOT MALLEABLE AT ANY TEMPERATURE
T -- 20 - 60 S -- 1 - 1.6T Y -- N/A	438.7 - 482.4	HIGH	LOW STRENGTH AND DUCTILITY
T -- 20 - 60 S -- 1 - 1.6T Y -- N/A	438.7 - 482.4	MEDIUM	LOW IMPACT STRENGTH
T -- 20 - 60 S -- 1 - 1.6T Y -- N/A	438.7 - 482.4	HIGH	COST
T -- 20 - 60 S -- 1 - 1.6T Y -- N/A	438.7 - 482.4	MEDIUM	NONE
T -- 40 - 100 S -- N/A Y -- 30 - 80	438.7 - 482.4	MEDIUM	THICKNESS LIMITED TO 3 INCHES
T -- 60 - 100 S -- N/A Y -- 40 - 90	438.7 - 482.4	MEDIUM	NONE
T -- 20 - 60 S -- 1 - 1.6T Y -- N/A	438.7 - 482.4	LOW	MUST BE REFINED TO BE USEFUL
T -- 20 - 60 S -- 1 - 1.6T Y -- N/A	438.7 - 482.4	MEDIUM	BRITTLE AND NOT MACHINABLE
T -- 34 - 54 S -- .83T Y -- 23 - 32	486.7 - 493.0	LOW	NOT COMMERCIALLY MADE IN THE USA

Table 4 Nonferrous Metals

MATERIAL	APPLICATION INFORMATION		
NONFERROUS METALS / TYPES	1. PRIMARY USE	2. PRIMARY ATTRIBUTE	3. MAJOR ALLOYING ELEMENT (PERCENT)
1. ALUMINUM	AUTOMATIC TRANSMISSIONS, CONTAINERS, SIDING, FRAME	DUCTILE, LIGHT WEIGHT, RESISTANT TO OXIDATION	AL -- 99.95
2. BRASS, LEADED	HAND RAILS, HINGES, MOLDINGS	CORROSION RESISTANT, MACHINABLE	Cu -- 64 Zn -- 35 Pb -- 1
3. BRASS, NON-LEADED	RADIATOR CORES	CORROSION RESISTANT	Cu -- 65 Zn -- 35
4. COPPER	PLUMBING GOODS, FITTINGS, ETC., TELEPHONE WIRE	EXCELLENT CONDUCTOR OF HEAT AND ELECTRICITY	Cu -- 99.95
5. INCONEL	ENGINE EXHAUST MANIFOLDS	HEAT RESISTANT	Ni -- 76 Cu --.25 Fe -- 8
6. LEAD	PIPES, BATTERIES, SOLDER, SHOT	LOW MELTING TEMPERATURE	Pb -- 99.95
7. MAGNESIUM	AIRCRAFT PARTS	SALT AND WATER RESISTANT, FORMABLE	Mg -- 99.95
8. MONEL	VALVES, PUMP PARTS, PROPELLER SHAFTS	CORROSION RESISTANT, STRONG AND TOUGH	Ni -- 66 Cu -- 29 Fe -- .9
9. NICKEL	ELECTRICAL CONTACT PARTS, FOOD AND DRUG HANDLING	CORROSION RESISTANT, GOOD ELECTRICAL CONDUCTIVITY AND HEAT TRANSFER	Ni -- 99.98
10. TIN	COATED STEEL CONTAINERS FOR FOOD	NONTOXIC	Sn -- 99.95
11. TITANIUM	HEAT EXCHANGERS	LIGHT WEIGHT, HEAT RESISTANT	Ti -- 99.95
12. ZINC	PROTECTIVE COATINGS FOR IRON AND STEEL	STRONG AND STABLE WHEN DIE CAST	Zn -- 99.95

APPLICATION INFORMATION			
4. STRENGTH T - TENSILE S - SHEAR Y - YIELD (THOUSANDS OF PSI)	5. WEIGHT PER CUBIC FOOT (LBS.)	6. MELTING POINT DEG. F.	7. DISADVANTAGE
T -- 19 - 83 S -- 8 - 48 Y -- 4 - 73	168.5	1220	LOW STRENGTH
T -- 32 - 88 S -- 29 - 46 Y -- 12 - 62	511 - 536	1660 - 1823	LOW DUCTILITY AND PLASTICITY
T -- 34 - 130 S -- 28 - 60 Y -- 10 - 65	511 - 536	1660 - 1823	-------
T -- 32 - 66 S -- 22 - 33 Y -- 10 - 53	554.7	1981	POOR CASTING QUALITY
T -- 70 - 185 S -- ------- Y -- 25 - 175	549	2651	NONE
T -- 2 - 5 S -- ------- Y -- -------	708	621	LOW STRENGTH
T -- 22 - 55 S -- 19 - 27 Y -- 20 - 44	109	1204	NONE
T -- 65 - 170 S -- ------- Y -- 25 - 160	549	2651	NONE
T -- 45 - 165 S -- ------- Y -- 10 - 155	549	2651	NONE
T -- 21 - 38 S -- 23 - 43 Y -- 15 - 18	454	449	MECHANICALLY WEAK
T -- 50 - 135 S -- ------- Y -- 40 - 120	280	3272	TENDENCY TO GALL
T -- 19 - 31 S -- ------- Y -- -------	439 - 446	788	BRITTLE

Table 5 Plastics

MATERIAL	APPLICATION INFORMATION		
PLASTICS	1. PRIMARY USE	2. PRIMARY ATTRIBUTE	3. CHEMICAL COMPOSITION
TYPES			
1. ABS	TELEPHONES, TOOL HANDLES, AUTOMOTIVE COMPONENTS	IMPACT AND ABRASION RESISTANT	THERMOPLASTIC
2. ACETAL	BEARINGS, CAMS, GEARS, IMPELLERS, ETC.	STRONG AND ABRASION RESISTANT	THERMOPLASTIC
3. ACRYLICS	LENSES, SIGNS, WINDOWS	OPTICAL QUALITY	THERMOPLASTIC
4. NYLON	COMBS, ZIPPERS, FASTENERS	SELF-LUBRICATING AND CHEMICAL RESISTANT	THERMOPLASTIC
5. POLYESTER	GEARS AND CAMS	LOW FRICTION AND ABRASION RESISTANT	THERMOSETTING PLASTIC
6. POLYETHYLENE	BOTTLES, GARBAGE CANS, LUGGAGE, TOYS	GOOD ELECTRICAL AND MECHANICAL PROPERTIES	THERMOPLASTIC
7. POLYPROPYLENE	TV CABINETS	GOOD ELECTRICAL AND MECHANICAL PROPERTIES	THERMOPLASTIC
8. POLYSTYRENE	FURNITURE PARTS, AUTOMOTIVE COMPONENTS	AVERAGE PROPERTIES	THERMOPLASTIC
9. POLYVINYL CHLORIDE	CABLE COATINGS, TUBING, RECORDS, BUILDING COMPONENTS	CAN BE MADE WITH A WIDE RANGE OF PROPERTIES	THERMOPLASTIC
10. POLYURETHANES	WHEELS AND CASTERS	HIGH STRENGTH	ELASTOMER
11. EPOXIES	ELECTRICAL COMPONENTS	DIMENSIONAL STABILITY AND ADHESIVE PROPERTIES	THERMOSETTING PLASTIC
12. PHENOLICS	BOATS, CHAIRS, AUTOMOTIVE BODIES, ETC.	HIGH HEAT, WATER, AND CHEMICAL RESISTANT	THERMOSETTING PLASTIC

APPLICATION INFORMATION			
4. STRENGTH T - TENSILE (THOUSANDS OF PSI)	5. WEIGHT PER CUBIC FOOT (LBS.)	6. RELATIVE COST	7. DISADVANTAGE
T -- 4 - 8	68	MEDIUM	COST OF DISPOSAL HIGH. PARTS ARE NOT BIODEGRADABLE
T -- 10	91	MEDIUM	COST OF DISPOSAL HIGH. PARTS ARE NOT BIODEGRADABLE
T -- 5.5 - 10	72.8	HIGH	COST
T -- 8 - 10	71.5	MEDIUM	COST OF DISPOSAL HIGH. PARTS ARE NOT BIODEGRADABLE
T -- 4 - 10	69	LOW	COST OF DISPOSAL HIGH. PARTS ARE NOT BIODEGRADABLE
T -- 4	62	MEDIUM	COST OF DISPOSAL HIGH. PARTS ARE NOT BIODEGRADABLE
T -- 3.4 - 5.3	58	MEDIUM	COST OF DISPOSAL HIGH. PARTS ARE NOT BIODEGRADABLE
T -- 5 - 9	68.25	MEDIUM	COST OF DISPOSAL HIGH. PARTS ARE NOT BIODEGRADABLE
T -- 4 - 10	——	MEDIUM	NOT HEAT RESISTANT
T -- 5	30	MEDIUM	COST OF DISPOSAL HIGH. PARTS ARE NOT BIODEGRADABLE
T -- 4 - 13	71.5	HIGH	COST
T -- 5 - 9	78	LOW	COST OF DISPOSAL HIGH. PARTS ARE NOT BIODEGRADABLE

Table 6 Steel

MATERIAL	APPLICATION INFORMATION		
STEEL / TYPES	1. PRIMARY USE	2. PRIMARY ATTRIBUTE	3. MAJOR ALLOYING ELEMENT (PERCENT)
1. CARBON	VARIOUS AUTOMOTIVE APPLICATIONS	FORMABILITY	C -- .08 - 1.03 Mn -- .25 - 1.65
2. MANGANESE	CARBURIZED PARTS	HARDENABILITY	C -- .28 - .48 Mn -- 1.60 - 1.90
3. NICKEL	GEARS, SHAFTS HEAT TREATED PARTS	TOUGHNESS	Ni -- 4.75 - 5.25
4. NICKEL-CHROMIUM	GEARS AND HEAT TREATED PARTS	TOUGHNESS AND WEAR RESISTANCE	Ni -- 1.10 - 1.40 Cr -- .55 - 1.75
5. MOLYBDENUM	BOLTS, SHAFTS	IMPROVED TENSILE STRENGTH	Mo -- .20 - .60
6. CHROMIUM MOLYBDENUM	AXLES AND OTHER SHAFTS	HARDENABILITY, TENSILE STRENGTH, AND WEAR RESISTANCE	Cr -- .40 - 1.10 Mo -- .15 - .25
7. NICKEL CHROMIUM MOLYBDENUM	CAMS, GEARS	TOUGHNESS, HARDENABILITY, AND STRENGTH	Ni -- 1.65 - 2.00 Cr -- .40 - .90 Mo -- .20 - .30
8. NICKEL MOLYBDENUM	BOLTS GEARS	TOUGHNESS, STRENGTH	Ni -- 1.65 - 2.00 Mo -- .20 - .30
9. CHROMIUM	BALL BEARING BALLS AND RACES	HARDENABILITY, ABRASION RESISTANCE	Cr -- .30 - 1.10
10. CHROMIUM VANADIUM	FORGINGS	IMPROVED STRENGTH, HARDENABILITY, AND TOUGHNESS	Cr -- .50 - 1.10 V -- .10 - .15
11. SILICON MANGANESE	LEAF SPRINGS	HIGH STRENGTH, DUCTILITY	Si -- .20 - .35 Mn -- .45 -.65
12. STAINLESS	AIRCRAFT AND FOOD PROCESSING APPLICATIONS	CORROSION RESISTANCE	Cr -- 16 - 26 Ni -- 3.5 - 37

APPLICATION INFORMATION			
4. STRENGTH T - TENSILE S - SHEAR Y - YIELD (THOUSANDS OF PSI)	5. WEIGHT PER CUBIC FOOT (LBS.)	6. RELATIVE COST	7. SAE SERIES NUMBERS
T -- 57 - 213 S -- .75T Y -- 20 - 170	489 - 491	LOW	10XX, 11XX 12XX, 15XX
T -- 90 - 162 S -- .75T Y -- 27 - 149	489 - 491	MEDIUM	13XX
T -- 88 - 190 S -- .75T Y -- 60 -155	489 - 491	MEDIUM	23XX, 25XX
T -- 93 - 188 Y -- .75T Y -- 62 - 162	489 - 491	MEDIUM	31XX, 32XX 33XX, 34XX
T -- 105 - 170 S -- .75T Y -- 60 - 114	489 - 491	MEDIUM	40XX, 44XX
T -- 81 - 179 S -- .75T Y -- 46 - 161	489 - 491	HIGH	41XX
T -- 109 - 220 S -- .75T Y -- 68 - 200	489 - 491	MEDIUM	43XX, 47XX, 81XX, 86XX-88XX, 93XX, 94XX-97XX, 98XX
T -- 98 - 192 S -- .75T Y -- 62 - 169	489 - 491	MEDIUM	46XX, 48XX
T -- 100 - 238 S -- .75T Y -- 81 - 228	489 - 491	HIGH	50XX, 51XX, 501XX, 511XX, 521XX
T -- 96 - 228 S -- .75T Y -- 69 - 206	489 - 491	HIGH	61XX
T -- 100 - 180 S -- .75T Y -- 62 - 169	489 - 491	MEDIUM	92XX
T -- 85 - 125 S -- .75T Y -- 30 - 95	489 - 491	HIGH	302XX, 303XX, 514XX, 515XX

Table 7 Tool Steel

MATERIAL	APPLICATION INFORMATION		
TOOL STEEL TYPES	1. PRIMARY USE	2. PRIMARY ATTRIBUTE	3. WORKING HARDNESS ROCKWELL C
1. WATER HARDENING	SHORT PRODUCTION RUN TOOLS	TOUGH AND ELASTIC STRUCTURE	50 - 64
2. SHOCK RESISTING	HOT SWAGING, DIES, BLACKSMITH TOOLS	SHOCK RESISTANT	45 - 60
3. COLD WORK OIL HARDENING	KEEN EDGE LOW SPEED TOOLING	STABLE HEAT TREATMENT WITH SUBSTANTIAL ABRASION RESISTANCE	57 - 64
4. COLD WORK AIR HARDENING	DIES, MEDIUM RUN	DIMENSIONAL CONTROL, WITH ABRASION AND WEAR RESISTANCE	57 - 62
5. COLD WORK HIGH CARBON HIGH CHROMIUM	DIES, LONG RUN	DIMENSIONAL CONTROL, WITH ABRASION AND WEAR RESISTANCE	54 - 61
6. HOT WORK CHROMIUM BASE	COLD HEADING, EXTRUSION, AND AIRCRAFT COMPONENTS	HEAT RESISTANT	38 - 55
7. HOT WORK TUNGSTEN BASE	DIE CASTING AND MOLDING DIES FOR BRASS	HEAT RESISTANT	36 - 54
8. HIGH SPEED STEEL TUNGSTEN BASE	CUTTING TOOLS	HIGH TEMPERATURE HARDNESS	60 - 66
9. HIGH SPEED STEEL MOLYBDENUM BASE	CUTTING TOOLS	HIGH TEMPERATURE HARDNESS	60 - 66
10. SPECIAL PURPOSE LOW ALLOY	BLANKING, FORMING AND TRIM DIES	TOUGHNESS	45 - 63
11.			
12.			

APPLICATION INFORMATION			
4. DEPTH OF HARDENING	5. WEIGHT PER CUBIC FOOT (LBS.)	6. MACHINABILITY	7. SAE SERIES NUMBER
SHALLOW	481	BEST	W108 - W110. W112. W209. W210. W310
MEDIUM	481	FAIR	S1. S2. S5
MEDIUM	481	GOOD	01. 02. 06
DEEP	481	FAIR	A2
DEEP	481	POOR	D2. D3. D5. D7
DEEP	481	FAIR	H11. H12. H13
DEEP	481	FAIR	H21
DEEP	481	FAIR	T1. T2. T4. T5. T8
DEEP	481	FAIR	M1. M2. M3. M4
MEDIUM	481	GOOD	L6. L7

Table 8 Wood

MATERIAL	APPLICATION INFORMATION		
WOOD TYPES	1. PRIMARY USE	2. PRIMARY ATTRIBUTE	3. RELATIVE DENSITY
1. ASH, WHITE	SPORTS EQUIPMENT, SKIS, BB BATS, TOBOGGANS, AND FURNITURE	STRONG, HARD, AND HOLDS SHAPE AFTER FORMING	HARD
2. BASSWOOD	DRAWING BOARDS AND SMALL MOLDINGS	NO WARPING AND NO GRAIN MARKS	SOFT
3. BIRCH, YELLOW	PANELS, DOORS, AND TABLE TOPS	RARE WARY GRAIN	HARD
4. CEDAR	CHESTS AND CLOSET PANELS	FINE GRAIN AND PLEASANTLY AROMATIC	SOFT
5. ELM	BENT PARTS OF FURNITURE	SHRINKS AND WARPS DURING CURING	MEDIUM
6. FIR, DOUGLAS	LUMBER AND PLYWOOD	HARD, HEAVY, AND VERY STIFF	SOFT
7. MAHOGANY	CABINETS	FINE GRAIN, UNIFORM TEXTURE, AND NATURAL COLOR	MEDIUM
8. MAPLE, HARD	FLOORING AND COLONIAL FURNITURE	GOOD WEAR AND SHOCK RESISTANCE	HARD
9. OAK, RED	FURNITURE, FLOORING AND TRIM	HARD, HEAVY, AND STRONG	HARD
10. OAK, WHITE	NATURALLY FINISHED CABINETS AND FURNITURE	FINE TEXTURE AND PROMINENT GRAIN	HARD
11. REDWOOD	FENCES AND OUTDOOR FURNITURE	RESISTANT TO DECAY	SOFT
12. WALNUT	CABINETS, FURNITURE VENEERS, AND GUNSTOCKS	BEAUTIFUL DARKWOOD EASILY WORKED	HARD

APPLICATION INFORMATION			
4. STRENGTH GENERAL	5. WEIGHT PER CUBIC FOOT (LBS) GREEN/AIRDRY	6. RELATIVE COST	7. DISADVANTAGE
GOOD	48/41	HIGH	HARD TO WORK
LOW	42/26	LOW	LONG STRINGY FIBERS
GOOD	57/44	MEDIUM	LOW RESISTANCE TO DECAY
LOW	37/33	HIGH	LOW RESISTANCE TO DECAY
MEDIUM	54/35	MEDIUM	SHRINKS AND WARPS
MEDIUM	46/27	LOW	NONE
MEDIUM	----	HIGH	COST
GOOD	50/38	MEDIUM	LOW RESISTANCE TO DECAY
GOOD	64/44	MEDIUM	HARD TO WORK
GOOD	63/47	MEDIUM	HARD TO WORK
MEDIUM	50/28	MEDIUM	NONE
GOOD	58/38	HIGH	COST

APPENDIX III

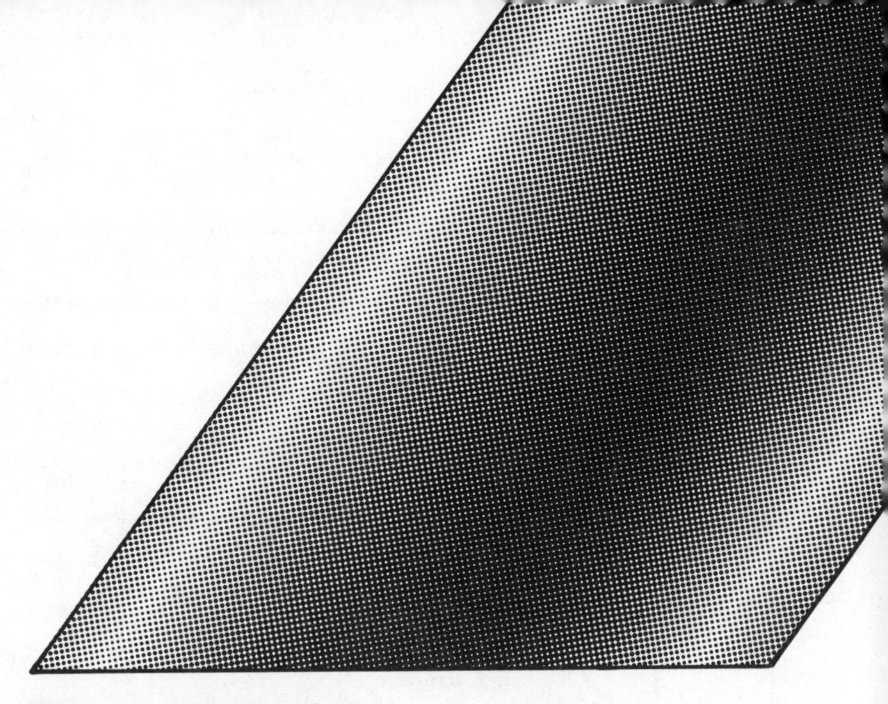

Processes

Table 1 Casting

PROCESS	APPLICATION INFORMATION		
CASTING TYPES	1. TYPICAL APPLICATIONS	2. SIZE RANGE	3. TYPICAL TOLERANCE AND NOMINAL SURFACE FINISH HELD
1. CENTRIFUGAL	A VARIATION ON THE SHELL MOLDING PROCESS FOR MORE INTRICATE PARTS	UNLIMITED	TOL. ±.030 IN./IN. MICROFINISH 250
2. DIE	ZINC, ALUMINUM, MAGNESIUM AND BRASS ALLOYS	UP TO 15 LBS.	TOL. ±.0015 IN./IN. MICROFINISH 63
3. DRY SAND	MOST FERROUS AND NONFERROUS ALLOYS	UNLIMITED	TOL. ±.030 IN./IN. MICROFINISH 500-1000
4. GREEN SAND	MOST FERROUS AND NONFERROUS ALLOYS	UNLIMITED	TOL. ±.030 IN./IN. MICROFINISH 63-125
5. GRAPHITE PERMANENT MOLD	ZINC AND ALUMINUM ALLOYS	UP TO 10 LBS.	TOL. ±.002 IN./IN. MICROFINISH 63-125
6. INVESTMENT	EXTREMELY COMPLEX SHAPES IN ANY METAL	UP TO 50 LBS.	TOL. ±.002 IN./IN. MICROFINISH 63-125
7. PERMANENT	ZINC, ALUMINUM, BRASS AND BRONZE	UP TO 150 LBS.	TOL. ±.002 IN./IN. MICROFINISH 63-125
8. PLASTER MOLD	PROTOTYPING IN ZINC, ALUMINUM, BRASS AND BRONZE	UP TO 20 LBS.	TOL. ±.002 IN./IN. MICROFINISH 63-125
9. SHELL MOLD	MOST FERROUS AND NONFERROUS ALLOYS	UP TO 50 LBS.	TOL. ±.008 IN./IN. MICROFINISH 500
10.			
11.			
12.			

APPLICATION INFORMATION			
4. TYPICAL PRODUCTION VOLUME	5. RELATIVE TOOLING COST	6. DISADVANTAGE TO USE	7. COMMENTS
UNLIMITED	HIGH	CASTINGS MUST BE MACHINED	CERAMIC MOLDS ARE SPUN AS METAL IS POURED TO AID IN MOLD FILL
VERY HIGH MILLIONS/YR.	HIGH	PARTS MAY CONTAIN A HIGH DEGREE OF POROSITY	METAL IS FORCED INTO CLOSED STEEL DIE UNDER PRESSURE
UNLIMITED	MEDIUM	CASTINGS MUST BE MACHINED	DRIED GREEN SAND GIVES GREATER MOLD STRENGTH
UNLIMITED	LOW	CASTINGS MUST BE MACHINED	GREEN SAND IS A MIXTURE OF SAND, CLAY, AND WATER. (IS REUSABLE)
500-20,000 PER YEAR	MEDIUM	LIMITED TO LOW PRODUCTION RUNS	LIKE PERMANENT MOLD CASTING EXCEPT CAVITIES ARE MACHINED IN GRAPHITE
LOW VOLUME ONE TO THOUSANDS	MEDIUM	ORIGINAL TOOLING COSTS ARE HIGH	CALLED LOST WAX OR PRECISION CASTING
5,000-50,000 PER YEAR	MEDIUM	NOT SUITABLE FOR HIGH MELTING TEMP. METALS (i.e. STEEL)	MOLD CAVITIES ARE IN METAL DIE BLOCKS
1-100 PER YEAR	VERY LOW	LIMITED TO NONFERROUS METALS	A PLASTER SLURRY IS POURED OVER A PATTERN
500-5,000 PER YEAR	MEDIUM	EXPENSIVE PATTERN AND EQUIPMENT REQUIRED	RESIN COATED SAND IS FORMED AGAINST HEATED METAL PATTERNS

Table 2 Cleaning

PROCESS	APPLICATION INFORMATION		
CLEANING TYPES	1. TYPICAL APPLICATIONS	2. REASON FOR SELECTION	3. OTHER BENEFITS
1. PETROLEUM SOLVENT	REMOVAL OF HEAVY SOIL OR RAPID LIGHT CLEANING	WATER BASED SOLVENTS WILL NOT CUT SOME TYPES OF SOILS	WILL LEAVE A LIGHT SOLVENT RESIDUE AS PROTECTION AGAINST RUST
2. EMULSIFIABLE SOLVENT	REMOVAL OF HEAVY OIL FILMS, WAXES, GREASES AND RUST PREVENTIVES	USED WITH SPRAY APPLICATION TO AID IN SOIL REMOVAL	WILL LEAVE A LIGHT SOLVENT RESIDUE AS PROTECTION AGAINST RUST
3. ELECTROLYTIC ALKALINE	CLEANING OF BRASS, NICKEL, LEAD AND NICKEL SILVER	ELECTRIC CURRENT RELEASES GAS BUBBLES WHICH SCRUB THE PART	NOT A BENEFIT: MAY PRODUCE HYDROGEN EMBRITTLEMENT IN FERROUS METALS
4. ALKALINE DESCALING	REMOVAL OF SOIL, PAINTS, OIL, SCALE AND OXIDES FROM ALUMINUM, ZINC, AND STEEL	ELIMINATES THE NEED FOR ACID DESCALING	NOT A BENEFIT: MAY PRODUCE HYDROGEN EMBRITTLEMENT IN FERROUS METALS
5. ACIDIC PICKLING	REMOVAL OF OXIDES, HEAT SCALE, STAINS AND DISCOLORATIONS	IMPROVED PAINT ADHESION AFTER PICKLING	CAN BE BLENDED WITH A WETTING AGENT TO REMOVE SOIL
6. ACIDIC SALT BATH DESCALING	REMOVAL OF SCALE AND OXIDES	VERY DIFFICULT CLEANING JOBS	NONE
7. STEAM	REMOVAL OF SOILS AND GREASES	CAN GET INTO OTHERWISE INACCESSIBLE AREAS	HIGHLY PORTABLE
8. FLAME	REMOVAL OF LOOSE MILL SCALE AND RUST	FAST, CHEAP, AND PORTABLE	WILL ALSO REMOVE OIL AND WATER
9. VAPOR DECREASING	REMOVAL OF SOILS, OILS AND GREASE	WILL CLEAN CREVICES AND CAPILLARIES OF METAL, PLASTIC, AND PORCELAIN	EXTREMELY LOW COST CLEANING
10. ULTRASONIC	REMOVAL OF CHIPS, LUBRICANTS, LIGHT OXIDES, ETC.	WILL CLEAN RECESSES OF COMPLEX PARTS	MAY BE AUTOMATED
11.			
12.			

APPLICATION INFORMATION			
4. TIME REQUIRED (MINUTES)	5. MACHINERY COST	6. DISADVANTAGE TO USE	7. COMMENTS
1- 15	MEDIUM	FUMES MAY BE TOXIC OR FLAMMABLE	USUALLY A PRECLEANING PROCESS
1- 15	MEDIUM	COST OF SOLVENT	ALLOWS THE USE OF WATER SOLVENT MIXTURES
.5 - 2	MEDIUM	A RINSE IS REQUIRED AFTER THE ALKALINE CLEANING	PARTS ARE IMMERSED IN A 140 - 180 F ALKALINE SOLUTION
.5 - 3	MEDIUM	NOT RECOMMENDED FOR USE ON STAINLESS, MAGNESIUM, COPPER OR BRASS	COMPLETED IN TANKS, BARREL, WITH SPRAY, OR IMMERSION WITH CURRENT
1 - 30	MEDIUM	PICKLING MUST BE FOLLOWED WITH A RINSE TO STOP FURTHER ATTACK	VENTING MAY BE REQUIRED
.5 - 5	HIGH	MAY REQUIRE A RINSE AND A NEUTRALIZING DIP	PARTS ARE IMMERSED IN A 400 - 1000 F SALT BATH
DEPENDS UPON SIZE OF PART	LOW	PAINT DOES NOT ADHERE WELL TO STEAM CLEANED PARTS	CAN BE MANUAL OR AUTOMATIC
DEPENDS UPON SIZE OF PART	LOW	VENTILATION REQUIRED	PROVIDES A WARM DRY SURFACE TO BE PAINTED
2 - 5	MEDIUM	WILL NOT REMOVE OXIDES, SCALES AND SOME INORGANIC FLUXES	NORMALLY CONVEYORIZED WITH PARTS IN BASKETS OR PANS
2 - 5	HIGH	THE SOLVENT SOLUTION MUST BE DEGASSED TAKING UP TO ONE HOUR	CLEANING CHEMICALS ARE ALSO REQUIRED (IMMERSION SOLUTION)

Table 3 Deburring

PROCESS	APPLICATION INFORMATION		
DEBURRING TYPES	1. TYPICAL APPLICATIONS	2. PREPARATORY PROCESS	3. SIZE BURR REMOVED AND NOMINAL SURFACE FINISH HELD
1. ABRASIVE JET	IMPACT ABRASION AND PEENING	VARIOUS MACHINING, STAMPING, CASTING, AND FORGING PROCESSES	BURR .005 MICROFINISH 40
2. BRUSH	EXTERNAL EDGES IRREGULARLY SHAPED PARTS	VARIOUS MACHINING, STAMPING, CASTING, AND FORGING PROCESSES	BURR .005 MICROFINISH --
3. CENTRIFUGAL BARREL	SUBSTANTIAL PARTS	VARIOUS MACHINING, STAMPING, CASTING, AND FORGING PROCESSES	BURR .005 MICROFINISH 40
4. CHEMICAL	SMALL PARTS	VARIOUS MACHINING, STAMPING, CASTING, AND FORGING PROCESSES	BURR .005 MICROFINISH 35
5. ELECTRO-CHEMICAL	SMALL PARTS	VARIOUS MACHINING, STAMPING, CASTING, AND FORGING PROCESSES	BURR .005 MICROFINISH 35
6. HAND	LOW VOLUME IRREGULARLY SHAPED PARTS	VARIOUS MACHINING, STAMPING, CASTING, AND FORGING PROCESSES	BURR .020 MICROFINISH 63
7. MECHANICAL	CHAMFERING KNIVES, GRINDING WHEELS	VARIOUS MACHINING, STAMPING, CASTING, AND FORGING PROCESSES	BURR .050 MICROFINISH --
8. SANDING	DISKS, FLAT PARTS, AND CONTOURED PARTS	VARIOUS MACHINING, STAMPING, CASTING, AND FORGING PROCESSES	BURR .005 MICROFINISH 35
9. SPINDLE	PART WHERE RADIUSED EDGES ARE DESIRABLE	VARIOUS MACHINING, STAMPING, CASTING, AND FORGING PROCESSES	BURR .005 MICROFINISH 40
10. THERMAL	PARTS WITH MANY BURRS	VARIOUS MACHINING, STAMPING, CASTING, AND FORGING PROCESSES	BURR .005 MICROFINISH --
11. VIBRATORY	DEBURRING, CLEANING AND POLISHING	VARIOUS MACHINING, STAMPING, CASTING, AND FORGING PROCESSES	BURR .005 MICROFINISH 50
12. WATER JET	CUTTING BURRS AND FLASH	VARIOUS MACHINING, STAMPING, CASTING, AND FORGING PROCESSES	BURR .005 MICROFINISH 40

APPLICATION INFORMATION			
4. TYPICAL PRODUCTION VOLUME PER HOUR	5. RELATIVE TOOLING COST	6. DISADVANTAGE TO USE	7. COMMENTS
12 - 120	MEDIUM	OPERATOR'S VIEW OBSTRUCTED, CLOTHES AND FEET WET	A FAMILY OF PROCESSES BOTH WET AND DRY
240	LOW	HIGH DIRECT LABOR CONTENT	MOTORIZED ROTATING BRUSHES
BATCH PROCESS 5 - 30 MINS	LOW	WILL RADIUS EDGES	25 - 50 TIMES FASTER THAN BARREL TUMBLING
BATCH PROCESS 5 - 30 MINS	LOW	CHEMICAL MUST BE NEUTRALIZED	NONE
120	LOW	CHEMICAL MUST BE NEUTRALIZED	NONE
240	LOW	HIGH DIRECT LABOR CONTENT	NONE
120 - 400	LOW	NONE	INCORPORATION IN STD. MACH. PROCESS ELIMINATES SECONDARY DEBURRING
600	LOW	THE PROCESS ITSELF PRODUCES A BURR	NONE
30 - 60	MEDIUM	HIGH DIRECT LABOR CONTENT	WORKPIECE IS FASTENED TO THE END OF A ROTATING SHAFT
80	MEDIUM	NONE	IGNITION OF NATURAL GAS VAPORIZES BURR
120	MEDIUM	NONE	SIMILAR TO BARREL TUMBLING
120	MEDIUM	OPERATOR'S VIEW OBSTRUCTED, CLOTHES AND FEET WET	HIGH VELOCITY WATER JET

Table 4 Gaging

PROCESS	APPLICATION INFORMATION		
GAGING TYPES	1. TYPICAL APPLICATIONS	2. PREPARATORY PROCESS	3. TYPICAL TOLERANCE GAGED
1. MICROMETERS AND VERNIERS	TOOLROOM AND OTHER LOW VOLUME MACHINED PARTS	VARIOUS FORMING AND MACHINING PROCESSES	±.001 IN.
2. FUNCTIONAL (GO NO-GO)	HOLE SIZES, SHAFT DIAMETERS, STEP HEIGHTS, ETC.	VARIOUS FORMING AND MACHINING PROCESSES	±.002 IN.
3. EXTERNAL COMPARATOR AND MASTER DISC	GROUND PARTS	VARIOUS FORMING AND MACHINING PROCESSES	±.0002 IN.
4. BENCH CENTER	CHECKING CIRCULAR AND TOTAL RUNOUT	VARIOUS FORMING AND MACHINING PROCESSES	.001 IN. RUNOUT
5. AIR	CHECKING BORES	VARIOUS FORMING AND MACHINING PROCESSES	±.00025 IN.
6. ELECTRONIC	CHECKING MULTIPLE DIMENSIONS	VARIOUS FORMING AND MACHINING PROCESSES	±.00005 IN.
7. INDICATOR	VARIOUS PRODUCTION APPLICATIONS	VARIOUS FORMING AND MACHINING PROCESSES	±.0005 IN.
8. PROFILOMETER	SURFACE ROUGHNESS	VARIOUS FORMING AND MACHINING PROCESSES	SURFACE TEXTURE
9. OPTICAL COMPARATOR	PROFILES	VARIOUS FORMING AND MACHINING PROCESSES	±.001 IN.
10. COORDINATE MEASURING MACHINE	LAYING OUT DIMENSIONS OF TOOLS AND GAGES	VARIOUS FORMING AND MACHINING PROCESSES	±.00005 IN.
11. IN-PROCESS CONTROL	CHECKING PARTS DURING GRINDING (CONTROLS MACHINE)	VARIOUS FORMING AND MACHINING PROCESSES	±.0002 IN.
12. VISION	CHECK FOR ABSENCE OR PRESENCE OF PART FEATURES	VARIOUS FORMING AND MACHINING PROCESSES	SEE TYPICAL APPLICATIONS

APPLICATION INFORMATION			
4. TYPICAL PRODUCTION VOLUME	5. RELATIVE TOOLING COST	6. DISADVANTAGE TO USE	7. COMMENTS
NOT A PRODUCTION GAGING PROCESS	LOW	EASY TO MISREAD	REQUIRES SKILLED LABOR
LOW-HIGH	LOW	DOES NOT GIVE PRECISE DIMENSIONAL INFORMATION	REQUIRES NO MASTER
LOW-HIGH	MEDIUM	NONE	KNOWN AS A SHEFFIELD SHADOW GAGE
LOW	LOW	ONLY GOOD FOR PARTS WITH CENTERS OR BORES	ALSO USED IN THE INSPECTION LAB
LOW-HIGH	MEDIUM	MUST BE RECALIBRATED EVERY TWO HOURS	REQUIRES A MASTER
HIGH	HIGH	COST	REQUIRES A MASTER
MEDIUM	LOW-MEDIUM	NONE	REQUIRES A MASTER
LOW	MEDIUM	NONE	TOO DELICATE TO BE STATIONED ON THE PRODUCTION FLOOR
LOW	MEDIUM	PART SIZE LIMITED	NEEDS A PLASTIC TEMPLATE FOR PRODUCTION USE
TOOL AND GAGE INSPECTION. ALSO LOW PRODUCTION	HIGH	REQUIRES SKILLED LABOR	SHOULD BE IN AN AIR CONDITIONED LAB
HIGH	MEDIUM	EASILY DAMAGED IF PARTS ARE MISLOADED	PARTS MUST STILL BE INSPECTED
HIGH	HIGH	DOES NOT CHECK DIMENSIONS	STILL BEING PERFECTED

Table 5 Gear Manufacturing

PROCESS	APPLICATION INFORMATION		
GEAR MANUFACTURING TYPES	1. TYPICAL APPLICATIONS	2. PREPARATORY PROCESS	3. TYPICAL TOLERANCE AND NOMINAL SURFACE FINISH HELD
1. BROACHING	INTERNAL SPLINES	GENERAL MACHINING	TOL. .002 IN. MICROFINISH 63
2. FORMING	DIECASTING, WARM FORGING, STAMPING POWDERED METAL, AND PLASTIC MOLDING	VARIES WITH FORMING PROCESS	VARIES WITH FORMING PROCESS (SEE OTHER PROCESS CHARTS)
3. G TRAC	SPUR AND HELICAL GEARS	MACHINING OR STAMPING	TOL. .002 IN. MICROFINISH 63
4. HOBBING	GENERATING EXTERNAL SPUR AND HELICAL GEARS	GENERAL MACHINING	TOL. .010 IN. MICROFINISH 94
5. MILLING	A VARIETY OF BEVEL AND HYPOID GEARS	CASTING, FORGING, COLD OR HOT ROLLING, AND GENERAL MACHINING	TOL. .010 IN. MICROFINISH 94
6. ROLLING	EXTERNAL SPLINES AND WORMS	FORGING, COLD OR HOT ROLLING	TOL. .001 IN. MICROFINISH 16
7. SHAPING	GENERATING INTERNAL AND EXTERNAL GEAR TEETH	GENERAL MACHINING	TOL. .005 IN. MICROFINISH 63
8. GRINDING	AIRCRAFT AND HIGH PITCH LINE VELOCITY GEARS	SEE OTHER GEAR MANUFACTURING PROCESSES #1-7	TOL. .0005 IN. MICROFINISH 8-16
9. HONING	FINISHING OF HARDENED GEARS	SEE OTHER GEAR MANUFACTURING PROCESSES #1-7	LESSENS OR IMPROVES SOUND LEVEL OF MATING GEARS
10. LAPPING	FINISHING OF HARDENED GEARS	SEE OTHER GEAR MANUFACTURING PROCESSES #1-7	CORRECTS HEAT TREATMENT DISTORTION
11. ROLLING (FINISHING)	EXTERNAL SPLINES AND WORMS USED IN THE AUTO INDUSTRY	SEE OTHER GEAR MANUFACTURING PROCESSES #1-7	ADDS STRENGTH TO THE GEAR TEETH THROUGH WORK HARDENING
12. SHAVING	CUTTING OPERATION USED ON SOFT GEARS	SEE OTHER GEAR MANUFACTURING PROCESSES #1-7	CORRECTS MINOR ERRORS IN TOOTH SPACING

APPLICATION INFORMATION			
4. TYPICAL PRODUCTION VOLUME	5. RELATIVE TOOLING COST	6. DISADVANTAGE TO USE	7. COMMENTS
150/HOUR	HIGH INITIAL COST	REQUIRES A SPECIAL MACHINE AND TOOLING	EXTERNAL SHAPES COMPLETED USING A PROCESS CALLED POT BROACHING
300/HOUR	HIGH	SOME FORMED GEARS LACK STRENGTH. OTHERS LACK DIMENSIONAL ACCURACY	THIS IS REALLY A FAMILY OF VERY DIFFERENT PROCESSES
180-600/HOUR	HIGH	REQUIRES A SPECIAL MACHINE	SPECIFICALLY FOR EXTERNAL CUTTING OF STACKABLE FLAT GEARS
75/HOUR	HIGH	REQUIRES A SPECIAL MACHINE	CAN BE SPED UP WITH THE USE OF MULTIPLE THREAD HOBS
40/HOUR	MEDIUM	SLOW PRODUCTION	CAN BE COMPLETED ON A STD. MILLING MACHINE, HOB OR SPECIAL GEAR GENERATOR
500-1200/HOUR	HIGH	REQUIRES A SPECIAL MACHINE	CYLINDRICAL OR WEDGE SHAPED DIES USED
50/HOUR	HIGH	REQUIRES A SPECIAL MACHINE	CAN CUT TO WITHIN .100 IN. OF A SHOULDER
6/HOUR	HIGH	IT IS AN EXPENSIVE SECONDARY PROCESS	CAN PRODUCE AGMA CLASS #12-14 GEAR
60/HOUR	LOW	IT IS AN EXPENSIVE SECONDARY PROCESS	CAN PRODUCE AGMA CLASS #10-12 GEAR
60/HOUR	LOW	IT IS AN EXPENSIVE SECONDARY PROCESS	CAN PRODUCE AGMA CLASS #10-12 GEAR
1000/HOUR	MEDIUM	IT IS AN EXPENSIVE SECONDARY PROCESS	CAN PRODUCE AGMA CLASS #10-12 GEAR
60/HOUR	LOW	IT IS AN EXPENSIVE SECONDARY PROCESS	CAN PRODUCE AGMA CLASS #10-12 GEAR

Table 6 Heat Treatment

PROCESS	APPLICATION INFORMATION		
HEAT TREATMENT / TYPES	1. TYPICAL APPLICATIONS	2. PREPARATORY PROCESS	3. CASE DEPTH
1. CARBURIZING (CASE HARDENING)	LOW CARBON STEEL	VARIOUS MACHINING OPERATIONS	.035 - .062 INCH
2. CYANIDING (CASE HARDENING)	SMALL PARTS	VARIOUS MACHINING OPERATIONS	.010 INCH
3. CARBONITRIDING (CASE HARDENING)	PLAIN CARBON AND ALLOY STEELS	VARIOUS MACHINING OPERATIONS	.003 - .030 INCH
4. NITRIDING (CASE HARDENING)	AEROSPACE PARTS	VARIOUS MACHINING OPERATIONS	.010 - .030 INCH
5. FLAME HARDENING (SURFACE HARDENING)	URGENTLY NEEDED TOOLING	VARIOUS MACHINING OPERATIONS	.030 - .250 INCH
6. INDUCTION HARDENING (SURFACE HARDENING)	SUPERFICIAL SURFACE HARDENING OF PRODUCTION PARTS	VARIOUS MACHINING OPERATIONS	.010 - .030 INCH
7. THROUGH HARDENING	CUTTING TOOLS	VARIOUS MACHINING OPERATIONS	THROUGH
8. ANNEALING (SOFTENING TREATMENT)	TO IMPROVE MACHINABILITY AND COLD WORKING PROPERTIES	———	———
9. NORMALIZING (SOFTENING TREATMENT)	HOT WORKED STEEL PARTS	ELIMINATE DISTORTION DUE TO HEAT TREATMENT	———
10. STRESS RELIEVING (SOFTENING TREATMENT)	ELIMINATING STRESSES INTRODUCED INTO PARTS THROUGH A VARIETY OF PROCESSES	VARIOUS CASTING, MACHINING, FORMING, AND WELDING OPERATIONS	———
11.			
12.			

APPLICATION INFORMATION			
4. TYPICAL PRODUCTION VOLUME	5. RELATIVE COST	6. DISADVANTAGE TO USE	7. COMMENTS
1000/HOUR	MEDIUM	VARIES WITH SPECIFIC CARBURIZING TECHNIQUE	CARBON IS ADDED TO THE OUTER LAYER OF THE PART
1000/HOUR	HIGH	SALT BATH IS TOXIC AND ALSO EXPLOSIVE IF PARTS ARE WET	A BATCH PROCESS
1000/HOUR	MEDIUM	REQUIRES A CONTROLLED FURNACE ATMOSHPERE	ALSO CALLED CYANIDING, NICARBING, AND NITROCARBURIZING
1000/HOUR	HIGH	VERY SLOW 20 - 100 HOUR CYCLE TIME	REQUIRES VERY SLOW COOLING CYCLE
5/HOUR	HIGH	HARD TO CONTROL	HEATED BY OXYACETYLENE TORCH
200/HOUR	HIGH	NONE, GIVES LOCALIZED HARDENING OF LARGE OR COMPLEX PARTS	HEATED BY AN ELECTROMAGNETIC INDUCTION COIL
500/HOUR	MEDIUM	TOOLS MAY CRACK IF NOT DESIGNED PROPERLY	ALSO CALED TEMPERING, QUENCHING, AUSTEMPERING, AND MARTEMPERING
1000/HOUR	MEDIUM	IT'S AN ADDED OPERATION	HEAT AND HOLD AT HIGH SUBCRITICAL TEMPERATURES
1000/HOUR	MEDIUM	IT'S AN ADDED OPERATION	NORMALIZING MAY SOFTEN, HARDEN, OR STRESS RELIEVE STEEL
1000/HOUR	LOW	IT'S AN ADDED OPERATION	HEAT STEEL TO 1100 - 1300 F

Table 7 Machining (Nontraditional)

PROCESS	APPLICATION INFORMATION		
MACHINING (NONTRADITIONAL) TYPES	1. TYPICAL APPLICATIONS	2. PREPARATORY PROCESS	3. TYPICAL TOLERANCE AND NOMINAL SURFACE FINISH HELD
1. CHEMICAL MILLING	REMOVING LOCALIZED AREAS OR METAL ON IRREGULARLY SHAPED PARTS	USUALLY CLEAN AND DRY CASTING OR FORGING	TOL. ±.0025 IN. MICROFINISH 30
2. ELECTRICAL DISCHARGE GRINDING	GRINDING THIN AND/OR BRITTLE MATERIALS	NONE	TOL. ±.0001 IN. MICROFINISH 2-4
3. ELECTRICAL DISCHARGE MACHINING	MACHINING OF ELECTRICALLY CONDUCTIVE MATERIALS	NONE	TOL. ±.0001 IN. MICROFINISH 2-4
4. ELECTRO-CHEMICAL GRINDING	HARD AND TOUGH MATERIALS	NONE	TOL. ±.001 IN. MICROFINISH 8-12
5. ELECTRO-CHEMICAL HONING	HARD AND TOUGH MATERIALS	NONE	DIAMETERS TOL. ±.0005 IN. MICROFINISH 8-32
6. ELECTRO-CHEMICAL MACHINING	INTERNAL AND EXTERNAL SHAPES IN HARD MATERIALS	NONE	TOL. ±.001 IN. MICROFINISH 8-12
7. ELECTRON BEAM	DRILLING, SLOTTING, AND PERFORATION IN ALL MATERIALS EXCEPT DIAMOND	CLEAN AND DRY	TOL. ±.001 IN. MICROFINISH 40
8. HYDRODYNAMIC	SLITTING AND CONTOUR CUTTING OF NONMETALLIC MATERIALS	NONE	NOT APPLICABLE
9. LASER BEAM	CUT SMALL HOLES, TRIM THIN MATERIALS, AND CUT DIAMOND	CLEAN AND DRY	TOL. ±.0001 IN. MICROFINISH 2-10
10. PHOTO CHEMICAL	THIN GAGE STAMPINGS, FLAT AND COMPLEX	CLEAN AND DRY	TOL. ±.00025 IN. MICROFINISH 4-8
11. PLASMA ARC	CUTTING, GROOVING, GOUGING ELECTRICALLY CONDUCTIVE MATERIALS	NONE	TOL. ±.001 IN. MICROFINISH 32
12. ULTRASONIC	IRREGULAR SHAPES AND BLIND HOLES IN BRITTLE MATERIALS	NONE	TOL. ±.001 IN. MICROFINISH 60

APPLICATION INFORMATION			
4. TYPICAL PRODUCTION VOLUME	5. RELATIVE TOOLING COST	6. DISADVANTAGE TO USE	7. COMMENTS
VARIABLE	LOW	REQUIRES MASKING	CHEMICAL REMOVAL OF METAL
MATERIAL REMOVAL RATE UP TO 15 CUBIC IN./HOUR	LOW	DIELECTRIC FLUID MUST BE FILTERED	SPARK MACHINING NOT AFFECTED BY MATERIAL HARDNESS
MATERIAL REMOVAL RATE UP TO 15 CUBIC IN./HOUR	LOW	DIELECTRIC FLUID MUST BE FILTERED	SPARK MACHINING NOT AFFECTED BY MATERIAL HARDNESS
MATERIAL REMOVAL RATE .010 CUBIC IN./MIN.	LOW	NONE	MATERIAL IS REMOVED BY ELECTROCHEMICAL ATTACK
MATERIAL REMOVAL RATE .010 CUBIC IN./MIN.	LOW	NONE	CAN BE TIED INTO AUTOMATIC GAGING
MATERIAL REMOVAL RATE .010 CUBIC IN./MIN.	LOW	NONE	MATERIAL IS REMOVED BY ELECTROCHEMICAL ATTACK
PENETRATION RATE .010 IN./SEC.	HIGH	PROCESS PERFORMED IN A VACUUM	LOCALLY MELTS AND EVAPORATES MATERIAL
CUTTING SPEEDS UP TO 75 FPM	LOW (HIGHER WITH NC CONTROLS)	PRIMARILY FOR THIN MATERIALS	HIGH PRESSURE WATER STREAM WITH ADDITIVES
MATERIAL REMOVAL RATE .0005 CUBIC IN./HOUR	HIGH	SAFETY CONCERNS	LASER - LIGHT AMPLIFICATION BY STIMULATED EMISSION OR RADIATION
ONLY LIMITED BY NUMBER OF IMAGE MASKS	LOW	REQUIRES SPECIAL MASKING OF PHOTORESIST	MATERIALS .0001-.050 IN. THICK
VARIABLE	HIGH	NONE	MAXIMUM DEPTH OF CUT .375 IN.
PENETRATION RATE .008-.150 IN./MIN.	MEDIUM	CAN HAVE A TAPER IN HOLE .005 IN./IN.	VIBRATING TOOL WITH ABRASIVE SLURRY

Table 8 Machining (Traditional)

PROCESS	APPLICATION INFORMATION		
MACHINING (TRADITIONAL) / TYPES	1. TYPICAL APPLICATIONS	2. PREPARATORY PROCESS	3. TYPICAL TOLERANCE AND NOMINAL SURFACE FINISH HELD
1. BORING	TRUING AND SIZING INTERNAL DIAMETER	DRILLING	TOL. ±.001 IN. MICROFINISH - 63
2. BROACHING	RAPIDLY CUTTING INTERNAL AND EXTERNAL GEAR TYPE SHAPES	DRILLING	TOL. ±.002 IN. MICROFINISH - 63
3. COUNTERBORING AND SPOTFACING	CUTTING AN AREA FOR A BOLT HEAD TO SIT	DRILLING	TOL. ±.005 IN. MICROFINISH - 94
4. DRILLING	MAKING HOLE IN SOLID MATERIAL	MILLING, FACING, CORING OR NONE	TOL. $^{+.005}_{-.002}$ IN. MICROFINISH - 125
5. GRINDING	ACCURATELY REMOVING SMALL AMOUNTS OF MATERIAL	MILLING, DRILLING OR BORING	TOL. ±.001 IN. MICROFINISH - 16
6. HOBBING	CUTTING EXTERNAL GEAR TEETH	TURNING	TOL. ±.010 IN. MICROFINISH - 94
7. MILLING	CREATING A DATUM OR LOCATING SURFACE	CASTING, FORGING, COLD OR HOT ROLLING	TOL. ±.010 IN. MICROFINISH - 94
8. REAMING	ACCURATELY SIZING A DRILLED HOLE	DRILLING, DRILLING AND BORING, OR CORING	TOL. ±.001 IN. MICROFINISH - 63
9. SAWING	TURNING BAR STOCK INTO SLUGS	NONE	TOL. ±.050 IN. MICROFINISH - 250
10. SHAPING (GEAR)	CUTTING INTERNAL AND EXTERNAL GEAR TEETH	TURNING	TOL. ±.005 IN. MICROFINISH - 63
11. THREADING	CUTTING INTERNAL AND EXTERNAL THREADS	DRILLING FOR INTERNAL THREADS, TURNING FOR EXTERNAL THREADS	PITCH LINE TOL. ±.003 IN. MICROFINISH - 125
12. TURNING	TRUING AND SIZING EXTERNAL DIAMETERS	CHAMFERING, FACING, OR NONE	TOL. ±.005 IN. MICROFINISH - 94

APPLICATION INFORMATION			
4. TYPICAL PRODUCTION VOLUME (INCLUDES PART LOAD AND UNLOAD)	5. RELATIVE TOOLING COST	6. DISADVANTAGE TO USE	7. COMMENTS NOMINAL SFPM USING CARBIDE IN MACHINING LOW CARBON STEEL
60/HOUR	LOW	RELATIVELY SLOW OPERATION	350
150/HOUR	HIGH	REQUIRES A SPECIAL MACHINE AND TOOLING	25
60/HOUR	LOW	RELATIVELY SLOW OPERATION AND A WASTE OF MATERIAL	65 - 130
60/HOUR	LOW	RELATIVELY SLOW OPERATION AND A WASTE OF MATERIAL	80
75/HOUR	MEDIUM	EXPENSIVE SECONDARY OPERATION	WORK - 75 WHEEL - 5000
75/HOUR	HIGH	REQUIRES A SPECIAL MACHINE AND GEARING	200 WITH A HSS HOB
40/HOUR	MEDIUM	RELATIVELY SLOW OPERATION AND A WASTE OF MATERIAL	200 - 500
125/HOUR	LOW	SIZES HOLE ONLY. DOES NOT STRAIGHTEN HOLE	ROUGHING - 100 FINISHING - 50
60/HOUR	LOW	SAW CAN ANGLE OFF WHEN CUTTING THICK PARTS	100
50/HOUR	HIGH	REQUIRES A SPECIAL MACHINE AND GEARING	80
100/HOUR	MEDIUM	PROPER TAP AND DIE SELECTION REQUIRES SOME TRIAL AND ERROR	50
60/HOUR	MEDIUM	RELATIVELY SLOW OPERATION AND A WASTE OF MATERIAL	400

Table 9 Metal Forming

PROCESS	APPLICATION INFORMATION		
METAL FORMING TYPES	1. TYPICAL APPLICATIONS	2. PREPARATORY PROCESS	3. TYPICAL SIZE OF PART
1. COLD DRAWING	REDUCING THE SIZE OF METAL BAR, ROD, OR WIRE AND ALSO OTHER SHAPES	HEAT TREATMENT, SURFACE PREPARATION, AND POINTING	.001 IN. WIRE TO 12 IN. DIA. TUBES
2. HOT EXTRUSION	LONG PARTS OF UNIFORM CROSS SECTION	BILLET FORMATION AND HEATING	UNDER 80 SQ. INCHES CONFINED IN A 24 IN. CIRCLE
3. COLD/WARM EXTRUSION	FORMING SHELL CASINGS, FLANGED CYLINDERS,	MATERIAL MUST BE PLACED INTO SLUGS, ANNEALED AND CLEANED	2.5 IN. DIA. X 24 IN. LONG
4. UPSETTING (HEADING)	SCREWS, BOLTS, AND MANY OTHER PARTS	CUTTING WIRE INTO SPECIFIC BLANK LENGTHS	1.25 IN. DIAMETER X 2.5 IN. DIAMETERS IN LENGTH
5. SWAGING	TAPERING, POINTING, AND REDUCING EXTERNAL FORMS	CLEANING	.005 - 5 INCHES IN DIAMETER
6. POWDER METAL	REPLACES MANY PARTS THAT WERE CAST OR FORGED	MANUFACTURE OF METAL POWDER	35 LBS. OR LESS
7. EXPLOSIVE FORMING	LARGE AEROSPACE PARTS	REQUIRES A TUBE	UNLIMITED
8. ELECTRO-HYDRAULIC FORMING	FORMATION OF HOLLOW SHAPES	REQUIRES A TUBE	.25 - 60 INCHES IN DIAMETER
9. ELECTRO-MAGNETIC FORMING	ASSEMBLY OF TUBULAR MEMBERS	REQUIRES A TUBE	.10 - 72 INCHES IN DIAMETER
10. HIGH VELOCITY FORGING	FORGING OF COMPLEX SHAPES	CREATION OF A PREFORM	LIMITED BY RAM FORCE
11. PEEN FORMING	ADDS FATIGUE STRENGTH TO PARTS SUCH AS AXLES, GEARS, ETC.	ANY METAL WORKING PROCESS	UNLIMITED
12. ULTRASONIC ACTIVATED FORMING	AIDS IN VARIOUS OTHER FORMING PROCESSES	VARIOUS SHEET METAL PROCESSES	SEE OTHER FORMING PROCESSES

APPLICATION INFORMATION			
4. TYPICAL PRODUCTION VOLUME OR SPEED	5. RELATIVE TOOLING COST	6. DISADVANTAGE TO USE	7. COMMENTS
9 - 300 FT./MIN.	LOW	REDUCTION IS LIMITED TO 10 - 15% IN OVERALL AREA	PERFORMED AT ROOM TEMPERATURE
10 FT./MIN.	LOW	LOW PRODUCTION SPEEDS	HEATED BILLET IS FORGED THROUGH ONE OR MORE DIES
25 PARTS/HOUR	LOW	LIMITATIONS ON LENGTH TO DIAMETER RATIO	TAKES PLACE AT ROOM TEMPERATURE OR BELOW THE CRITICAL TEMP. OF THE METAL
35 - 900 PARTS/HOUR	HIGH	HIGH COST AND LIMITATION ON PART SIZE	PROVIDES LOW LABOR COST AND MATERIAL SAVINGS
100 - 1000 PARTS/HOUR	LOW	PARTS MUST BE SYMMETRICAL IN CROSS SECTION. ALSO A NOISY OPERATION	ROTARY IMPACT BLOWS ARE TRANSFERRED TO THE WORK
200 PARTS/HOUR	MEDIUM	PARTS ARE POROUS AND MAY NOT HAVE DESIRED STRENGTH	POWDER IS COMPACTED INTO SHAPES AND THEN SINTERED
4 PARTS/HOUR	LOW	REMOTE FACILITY AND SKILLED PERSONNEL REQUIRED	PUNCH IS REPLACED WITH AN EXPLOSIVE CHARGE
360 PARTS/HOUR	LOW	THE REQUIRED ENERGY AND A TRIGGERING DEVICE	ELECTRICAL ENERGY CAUSES EXPLOSION SENDING SHOCK WAVES THROUGH FLUID
12,000 PARTS/HOUR	MEDIUM	MATERIAL MUST BE A GOOD CONDUCTOR	MAGNETIC FIELD IS USED TO FORM PARTS
RAM SPEED 70FT./SEC. WITH 20,000 FT. LBS.	MEDIUM	CYCLE TIMES SLOWER AND TOOLING COSTS HIGHER THAN REGULAR FORGING	LIKE DROP HAMMER FORGING WITH THE AID OF AN EXPLOSION ABOVE THE RAM
60 PARTS/HOUR	VERY LOW	NONE	ALSO CALLED SHOT PEENING WHICH CAUSES LOCAL PLASTIC DEFORMATION
SEE OTHER FORMING PROCESSES	MEDIUM	POWER REQUIREMENTS	APPLIES HIGH FREQUENCY VIBRATIONS TO THE WORKPIECE

Table 10 Milling

PROCESS	APPLICATION INFORMATION		
MILLING TYPES	1. TYPICAL APPLICATIONS	2. PREPARATORY PROCESS	3. TYPICAL TOLERANCE AND NOMINAL SURFACE FINISH HELD
1. PERIPHERAL	END AND SLAB MILLING	CLEANING. MILLING. OR NONE	H.S.S. TOL. ± .005 IN. MICROFINISH - 125
2. FACE	INSERT MILLING. PRODUCTION APPLICATIONS	CLEANING. MILLING. OR NONE	CARBIDE TOL. ± .001 IN. MICROFINISH - 30
3. GANG	TWO OR MORE CUTTERS MOUNTED ON ONE ARBOR	CLEANING. MILLING. OR NONE	SEE ITEMS #1 AND 2
4. SINGLE PIECE	ALL MILLING OPERATIONS	CLEANING. MILLING. OR NONE	SEE ITEMS #1 AND 2
5. STRING	TWO OR MORE PARTS IN A ROW	CLEANING. MILLING. OR NONE	SEE ITEMS #1 AND 2
6. ABREAST	TWO PARTS SIDE BY SIDE	CLEANING. MILLING. OR NONE	SEE ITEMS #1 AND 2
7. PROGRESSIVE	MILLING TWO SIDES OF A PART WITH A SINGLE CUTTER	CLEANING. MILLING. OR NONE	SEE ITEMS #1 AND 2
8. BOX	FOUR SIDES OF A SQUARE OPENING. INTERNAL OR EXTERNAL	CLEANING. MILLING. OR NONE	SEE ITEMS #1 AND 2
9. RECIPROCAL	TWO FIXTURES MOUNTED AT EITHER END OF A RECIPROCATING TABLE	CLEANING. MILLING. OR NONE	SEE ITEMS #1 AND 2
10. TRANSFER BASE	UTILIZES AN INDEXING FIXTURE	CLEANING. MILLING. OR NONE	SEE ITEMS #1 AND 2
11. INDEX	GEAR TEETH AND OTHER EQUALLY SPACED FEATURES	CLEANING. MILLING. OR NONE	SEE ITEMS #1 AND 2
12. ROTARY	USES A ROTARY INDEX TABLE TO BRING PARTS UNDER CUTTER	CLEANING. MILLING. OR NONE	SEE ITEMS #1 AND 2

APPLICATION INFORMATION			
4. TYPICAL PRODUCTION VOLUME (INCLUDES PART LOAD AND UNLOAD)	5. RELATIVE TOOLING COST	6. DISADVANTAGE TO USE	7. NOMINAL SFPM MACHINING LOW CARBON STEEL
SEE ITEMS #3 -12	SEE ITEMS #3 -12	NOT AS FAST OR ACCURATE AS FACE MILLING	H.S.S. 25 - 150
SEE ITEMS #3 -12	SEE ITEMS #3 -12	CUT WIDTH LIMITED TO SIZE OF CUTTER (1.5 IN. MIN.)	CARBIDE 500
60/HOUR	LOW	MAY HAVE TO COMPROMISE SPEEDS AND FEEDS	SEE ITEMS #1 AND 2
71/HOUR	LOW	LOAD AND UNLOAD EXTERNAL TO CUTTING TIME	SEE ITEMS #1 AND 2
36/HOUR	MEDIUM	NONE	SEE ITEMS #1 AND 2
53/HOUR	MEDIUM	LONGER SET-UP TIMES	SEE ITEMS #1 AND 2
35/HOUR	HIGH	LIMITED HOURLY PRODUCTION CAPABILITY	SEE ITEMS #1 AND 2
35/HOUR	MEDIUM	LIMITED HOURLY PRODUCTION CAPABILITY	SEE ITEMS #1 AND 2
86/HOUR	HIGH	ONE PIECE IS CONVENTIONALLY MILLED. THE OTHER IS CLIMB MILLED	SEE ITEMS #1 AND 2
80/HOUR	HIGH	INDEX TIME MUST BE ADDED TO CUT TIME	SEE ITEMS #1 AND 2
5 - 10/HOUR	LOW	NONE	SEE ITEMS #1 AND 2
100/HOUR	HIGH	NONE	SEE ITEMS #1 AND 2

Table 11 Nondestructive Testing

PROCESS	APPLICATION INFORMATION		
NONDESTRUCTIVE TESTING	1. TYPICAL APPLICATIONS	2. PREPARATORY PROCESS	3. MEASURES OR DETECTS
TYPES			
1. EDDY CURRENT	METAL PARTS AND STRUCTURES	NONE	SUBSURFACE CRACKS, METALLURGICAL FACTORS, AND PLATING THICKNESS
2. FLUROSOCOPY	PARTICLES IN LIQUID FLOW	NONE	LEVEL OF FILL, FOREIGN OBJECTS, VOIDS, AND VARIATIONS
3. HOLOGRAPHY	MOST INDUSTRIAL MATERIALS	NONE	VOIDS, POROSITY, AND INCLUSIONS
4. LEAK TESTING	JOINTS - WELDED, BONDED, ETC.	DEGREASING	LEAKS - LIQUIDS AND GASES
5. MAGNETIC PARTICLE	FERROMAGNETIC MATERIALS	DEGREASING AND DRYING	SURFACE AND SUBSURFACE CRACKS, SEAMS, AND INCLUSIONS
6. DYE PENETRANTS	PARTS WITH NON-ABSORBING SURFACES	DEGREASING AND DRYING	SURFACE CRACKS AND VOIDS
7. RADIOGRAPHY	METALLIC AND NONMETALLIC ASSEMBLIES	NONE	INTERNAL DEFECTS
8. SONIC	METAL, PLYWOOD, AND HONEYCOMB CONSTRUCTION	NONE	BOND INTEGRITY AND DAMAGED CORES
9. ULTRASONIC	WELDS AND NONMETALLIC BONDED JOINTS	NONE	INTERNAL DEFECTS, CRACKS, POROSITY, AND LACK OF BONDING
10.			
11.			
12.			

APPLICATION INFORMATION			
4. RELATIVE PRODUCTION VOLUME	5. RELATIVE TOOLING COST	6. DISADVANTAGE TO USE	7. ADVANTAGES
LOW	HIGH	CAN ONLY SORT FOR ONE VARIABLE AT A TIME	DETECTS CRACKS NOT DETECTABLE BY RADIOGRAPHY
LOW	HIGH	SMALL VIEWING AREA	REAL TIME VIEWING, PERMANENT RECORD
LOW	HIGH	LASER REQUIRED	REAL TIME IMAGING
MEDIUM	LOW	SMEARED METAL MAY MASK DEFECT	HIGH SENSITIVITY
HIGH	LOW	ALIGNMEMT OF MAGNETIC FIELD CRITICAL	MAY BE PORTABLE
LOW	LOW	DEFECTS MUST BE EXPOSED TO THE SURFACE	EASILY INTERPRETED RESULTS
LOW	VERY HIGH	RADIATION HAZARD	REAL TIME VIEWING, PERMANENT RECORD
HIGH	MEDIUM	REFERENCE STANDARDS REQUIRED	ACCESS TO A SINGLE SURFACE REQUIRED
HIGH	MEDIUM	REFERENCE STANDARDS REQUIRED	HIGH SENSITIVITY

Table 12 Painting

PROCESS	APPLICATION INFORMATION		
PAINTING TYPES	1. TYPICAL APPLICATIONS	2. PREPARATORY PROCESS	3. SURFACE FINISH
1. COMPRESSED AIR SPRAY	CONVENTIONAL SPRAY PAINTING	NONE REQUIRED	FINE PAINT FINISH
2. AIRLESS SPRAY	SIMILAR TO CONVENTIONAL SPRAY PAINTING	NONE REQUIRED	FINE PAINT FINISH
3. HOT SPRAY	PAINTING WITH LACQUERS, ENAMELS, AND OIL BASE PAINT	HEATING OF THE PAINT TO 120-170 F	FINE PAINT FINISH
4. ELECTROSTATIC SPRAY	PRODUCTION PAINTING OF ELECTRICAL CONDUCTIVE PARTS	NONE REQUIRED	FINE PAINT FINISH
5. DIP COAT	PRODUCTS REQUIRING TOTAL COVERAGE	NONE REQUIRED	UNIFORM BUT NOT FINE
6. FLOW COAT	PRODUCTS REQUIRING TOTAL COVERAGE	NONE REQUIRED	UNIFORM BUT NOT FINE
7. CURTAIN COAT	APPLICATION OF PAINTS, VARNISHES, AND ADHESIVES TO LARGE FLAT SURFACES	NONE REQUIRED	UNIFORM BUT NOT FINE
8. ROLLER COAT	APPLICATION OF PAINTS, VARNISHES, AND ADHESIVES TO LARGE FLAT SURFACES	NONE REQUIRED	VERY FINE FINISHES
9. BARREL COAT	PAINTING OF SMALL PARTS (I.E. FASTENERS)	NONE REQUIRED	UNIFORM BUT NOT FINE
10. CENTRIFUGAL COAT	PAINTING OF SMALL PARTS (I.E. FASTENERS)	NONE REQUIRED	UNIFORM BUT NOT FINE
11. FLUIDIZED BED COAT	USING OF DRY POWDER	PARTS ARE PREHEATED IN AN OVEN	FINE PAINT FINISH
12. TRICHLOR-ETHYLENE FINISHING	PRODUCTION DEGREASING, PHOSPHATIZING AND BLACK LACQUER PAINTING OF PARTS	NONE REQUIRED	FINE PAINT FINISH

APPLICATION INFORMATION			
4. TYPICAL PRODUCTION VOLUME	5. RELATIVE TOOLING COST	6. DISADVANTAGE TO USE	7. COMMENTS
LOW	LOW	OVERSPRAY WASTES 40-70% OF PAINT. REQUIRES CAREFUL MIXTURE OF PAINT AND SOLVENT	PRODUCTION USE WOULD REQUIRE A SPRAY BOOTH
MEDIUM	MEDIUM	NOZZLE MUST BE CHANGED TO ADJUST SPRAY	HYDRAULIC PRESSURE ELIMINATES THE IMPURITIES OF COMPRESSED AIR
MEDIUM	MEDIUM	SOME PAINTS NOT IMPROVED BY HOT SPRAYING	PREHEATING OF THE PAINT GIVES UNIFORM VISCOSITY
HIGH	MEDIUM	BLIND HOLES AND DEEP POCKETS DO NOT GET FULL COVERAGE	80% UTILIZATION. UNLIKE ELECTRICAL CHARGES ATTRACT
HIGH	LOW	PARTS MUST HAVE A WAY TO DRAIN	AGITATION OF PAINT IN TANK REQUIRED
HIGH	MEDIUM	FILM THICKNESS WILL BE HEAVIER AT BOTTOM OF PART	LOW PRESSURE NOZZLE FLOODS CONVEYORIZE PARTS
HIGH	MEDIUM	NONE	MAY ALSO BE USED TO APPLY HOT WAX AND PLASTIC
HIGH	HIGH	HIGH INITIAL COST	UP TO 98% MATERIAL UTILIZATION
HIGH	LOW	PARTS MUST BE DURABLE. NOT SHARP OR INTERLOCKING	A METERED AMOUNT OF PAINT IS PLACED IN A BARREL WITH THE PARTS
HIGH	LOW	PARTS MUST BE DURABLE. NOT SHARP OR INTERLOCKING	PARTS ARE PLACED IN A WIRE BASKET AND SPUN
MEDIUM	MEDIUM	PARTS MUST BE POST HEATED TO FUSE PAINT	PARTS ARE IMMERSED AFTER PREHEATING
HIGH	HIGH	REQUIRES HIGH PRODUCTION TO JUSTIFY EQUIPMENT	MAY BE REFERRED TO AS THE DUPONT PROCESS

Table 13 Plating

PROCESS	APPLICATION INFORMATION		
PLATING	1. TYPICAL APPLICATIONS	2. PREPARATORY PROCESS	3. SURFACE FINISH
TYPES			
1. PHOSPHATE CONVERSION COATING	CORROSION PROTECTION AND PREPAINT TREATMENT	DEGREASING	GRAY METAL FLAKE TYPE OR MATTE FINISH
2. ANODIZING	A PREPAINT TYPE COATING OR PART DYING ON ALUMINUM AND MAGNESIUM	BUFFING, WIRE BRUSHING, ETC. AND CLEANING	BRIGHT, FLAT OR MATTE FINISH, VARIOUS COLORS
3. ELECTROPLATING	PLATING OF STEEL, BRASS AND ZINC (DIE CASTINGS)	SOAK CLEAN, COLD WATER RINSE, ELECTRO CLEAN, RINSE, ACID DIP, COLD WATER RINSE	NICKEL, COPPER, TIN, CHROMIUM, ZINC, LEAD, AND CADMIUM PLATE
4. MECHANICAL PLATING	STEEL PARTS UNDER ONE POUND IN WEIGHT	DESCALING AND SOIL REMOVAL	ZINC, CADMIUM, TIN, LEAD PLATE .2-1 MIL THICK
5. ELECTROLESS PLATING	DEPOSITING COPPER, GOLD, AND NICKEL PLATE TO A VARIETY OF PARTS	CLEANING	PRODUCES A UNIFORM COATING ON ALL PART SHAPES
6. THERMAL SPRAYING	REBUILDING OF WORN PARTS	DEGREASING AND GRIT BLASTING	100-400 MICROFINISH
7. HARD FACING	ADDING A HARD LAYER TO A PART OR REBUILDING OF WORN PARTS	CLEANING	DEPOSITS UP TO .125 IN. ARE COMMON
8. PORCELAIN ENAMELING	COATING OF A METAL SUBSTRATE WITH A LAYER OF GLASS (I.E. APPLIANCES)	CLEANING AND DEBURRING	COATING .002-.004 INCHES THICK
9. GALVANIZING	ZINC GALVANIZING OF STEEL PARTS	CLEANING AND FLUXING	COATING .002-.0035 INCHES THICK
10. HOT DIPPING	COVERING STEEL OR IRON SUBSTRATES WITH ALUMINUM, TIN, OR LEAD	CLEANING AND FLUXING	COATING .002-.0035 INCHES THICK
11.			
12.			

APPLICATION INFORMATION			
4. TYPICAL PRODUCTION VOLUME	5. RELATIVE TOOLING COST	6. DISADVANTAGE TO USE	7. COMMENTS
HIGH TEN SECONDS TO 30 MINUTES PER PART	MEDIUM	MAY SLIGHTLY ETCH THE PART SURFACE	ALSO USED FOR WEAR RESISTANCE, DECORATION, OR IDENTIFICATION
HIGH	MEDIUM	GOOD VENTILATION SYSTEM IS REQUIRED	COATING THICKNESS .0001-.003 IN.
HIGH WHEN CONVEYORIZED	HIGH	MANY PART DESIGNS CANNOT BE PLATED EVENLY	PLASTICS CAN ALSO BE PLATED
MEDIUM	HIGH	PART SIZE	GLASS BEADS ARE USED TO CARRY PLATING MATERIAL INTO HOLES ETC.
MEDIUM	MEDIUM	REQUIRES THE USE OF CORROSIVE, HAZARDOUS, AND TOXIC MATERIALS	IS A CHEMICAL REACTION ALSO CALLED AUTO-CATALYTIC DEPOSITION
LOW	MEDIUM	EMITS INTENSE ULTRAVIOLET RAYS AND NOISE	SPRAYING MOLTEN OR SEMIMOLTEN METAL ONTO A SUBSTRATE
LOW	LOW	NOT A PRODUCTION PROCESS	CREATES A METALLURGICAL BOND
HIGH	MEDIUM	DEFECTS MUST BE REPAIRED	MAY BE APPLIED BY DIPPING, FLOW COATING, SPRAYING AND ELECTRO METHODS
HIGH	MEDIUM	COATING CANNOT STAND UP TO MECHANICAL DAMAGE	ACCOMPLISHED BY HOT DIPPING
HIGH	MEDIUM	COATING CANNOT STAND UP TO MECHANICAL DAMAGE	ACCOMPLISHED BY HOT DIPPING

Table 14 Sheet Metal

PROCESS	APPLICATION INFORMATION		
SHEET METAL / TYPES	1. TYPICAL APPLICATIONS	2. PREPARATORY PROCESS	3. TYPICAL TOLERANCE
1. BLANKING AND PIERCING	ENTIRE OUTLINE IS CUT IN A SINGLE STROKE OF THE PRESS	SHEET METAL (SHEET OR COIL) CUT OR SLIT TO CORRECT SIZE	.003 - .035 IN. DEPENDING ON MATERIAL THICKNESS AND DIE OPENING
2. NOTCHING	METAL IS CUT FROM EDGE OF STRIP	SHEET METAL (SHEET OR COIL) CUT OR SLIT TO CORRECT SIZE	±.005 IN.
3. LANCING	COMBINATION CUTTING AND BENDING	SHEET METAL (SHEET OR COIL) CUT OR SLIT TO CORRECT SIZE	±.005 IN.
4. SHAVING	A SECONDARY OPERATION TO HOLD CLOSE TOLERANCE DIMENSIONS	SHEET METAL (SHEET OR COIL) CUT OR SLIT TO CORRECT SIZE	±.001 IN.
5. CUTOFF AND PARTING	REMOVING OR CUTTING A BLANK FROM A STRIP	SHEET METAL (SHEET OR COIL) CUT OR SLIT TO CORRECT SIZE	±.005 IN.
6. BENDING	STRENGTHENING OF SHEET METAL SHAPES	SHEET METAL (SHEET OR COIL) CUT OR SLIT TO CORRECT SIZE	±.005 IN.
7. SOLID AND PRESSURE PAD FORMING	FORMING CURVED AND INTRICATE SHAPES	SHEET METAL (SHEET OR COIL) CUT OR SLIT TO CORRECT SIZE	±.005 IN.
8. CURLING	PUTTING ROLLED EDGES ON SHEET METAL	SHEET METAL (SHEET OR COIL) CUT OR SLIT TO CORRECT SIZE	±.005 IN.
9. EMBOSSING AND COINING	PUTTING DETAIL (I.E. LETTERS) ON SHEET	SHEET METAL (SHEET OR COIL) CUT OR SLIT TO CORRECT SIZE	±.001 IN.
10. STRETCH FORMING	ADDING STRENGTH AND REDUCING WEIGHT TO CURVED FORMS	SHEET METAL (SHEET OR COIL) CUT OR SLIT TO CORRECT SIZE	±.005 IN.
11. DRAWING	MAKING CUP TYPE SHAPES FROM FLAT SHEET	SHEET METAL (SHEET OR COIL) CUT OR SLIT TO CORRECT SIZE	±.005 IN.
12. ROLL FORMING	AUTOMATIC PRODUCTION OF COMPLEX FORMS FROM FLAT SHEETS	SHEET METAL (SHEET OR COIL) CUT OR SLIT TO CORRECT SIZE	±.005 IN.

APPLICATION INFORMATION			
4. TYPICAL PRODUCTION VOLUME	5. RELATIVE TOOLING COST FOR INDIVIDUAL PROCESSES	6. DISADVANTAGE TO USE	7. COMMENTS
250/MINUTE	LOW	LEAVES SLIGHT TENSILE BURR ON ONE SIDE OF PART	STOCK THICKNESS FROM .030 - .625 IN.
250/MINUTE	LOW	LEAVES SLIGHT TENSILE BURR ON ONE SIDE OF PART	STOCK THICKNESS FROM .030 - .625 IN.
250/MINUTE	LOW	LEAVES SLIGHT TENSILE BURR ON ONE SIDE OF PART	STOCK THICKNESS FROM .030 - .625 IN.
250/MINUTE	LOW	LEAVES SLIGHT TENSILE BURR ON ONE SIDE OF PART	STOCK THICKNESS FROM .030 - .625 IN.
250/MINUTE	LOW	LEAVES SLIGHT TENSILE BURR ON ONE SIDE OF PART	STOCK THICKNESS FROM .030 - .625 IN.
DEPENDS ON MACHINE OR DIE TYPE BEING USED	LOW	NONE	CAN BE VEE, WIPE, OR ROTARY BENDING
DEPENDS ON MACHINE OR DIE TYPE BEING USED	MEDIUM	DOES NOT LEND ITSELF TO PROGRESSIVE DIE WORK	PRESSURE HELPS TO HOLD PART SO OTHER OPERATIONS CAN BE PERFORMED
DEPENDS ON MACHINE OR DIE TYPE BEING USED	LOW	METAL MAY SPLIT DURING ROLLOVER	ADD STRENGTH AND SAFETY TO PART
250/MINUTE	MEDIUM	DETAIL MUST BE VERY SHALLOW	ALSO USED FOR MAKING COINS
60/HOUR	MEDIUM	SHEET MAY NOT STRETCH UNIFORMLY	SHARP CONTOURS CANNOT BE MATED
720/HOUR	LOW	MAY REQUIRE MORE THAN ONE DRAW TO GET DESIRED SHAPE	ALSO REQUIRES A TRIMMING OPERATION
1,000,000 FT./YEAR (MINIMUM)	HIGH	CANNOT BE JUSTIFIED FOR LOWER VOLUMES	SOME SHAPES AND FORMS NOT POSSIBLE

Table 15 Threading

PROCESS	APPLICATION INFORMATION		
THREADING TYPES	1. TYPICAL APPLICATIONS	2. PREPARATORY PROCESS	3. TYPICAL TOLERANCE HELD AT THE PITCH DIAMETER
1. SINGLE POINT	EXTERNAL AND INTERNAL THREADING	TURNING AND BORING	±.001 IN.
2. DIE HEAD THREAD CHASING	EXTERNAL AND INTERNAL THREADING	TURNING AND BORING	±.001 IN.
3. THREAD ROLLING	THREADED FASTENERS	COLD HEADING	±.001 IN.
4. THREAD MILLING	MILLING OF WORM THREADS	TURNING AND BORING	±.002 IN.
5. THREAD GRINDING	EXTERNAL THREADS ON HIGHLY STRESSED PARTS	THREAD CUTTING OR TURNING	±.0002 IN.
6. CUT TAPPING	INTERNAL THREADS	TURNING AND BORING	±.0005 IN.
7. COLD FORM TAPPING	INTERNAL THREADS IN DUCTILE MATERIAL	TURNING AND BORING	±.0005 IN.
8. COLLAPSING TAP THREAD CHASING	INTERNAL THREADS ON LARGE DIAMETER PARTS	TURNING AND BORING	±.001 IN.
9.			
10.			
11.			
12.			

APPLICATION INFORMATION			
4. TYPICAL PRODUCTION VOLUME	5. RELATIVE TOOLING COST	6. DISADVANTAGE TO USE	7. COMMENTS
30/HOUR	LOW	LOW PRODUCTION RATE	COMPLETED ON A LATHE
60/HOUR	MEDIUM	MUST BE RESET BY HAND	MAY BE RUN ON A LATHE, DRILL PRESS, OR VERTICAL MILL
1000/MINUTE	MEDIUM	GENERALLY LIMITED TO 16 THREADS/IN. OR FINER	REQUIRES A SPECIAL MACHINE
20/HOUR	LOW	LOW PRODUCTION RATE	COMPLETED ON A MILLING MACHINE
150/HOUR	MEDIUM	MAINTENANCE OF THE ROOT WIDTH IS A PROBLEM	COMPLETED ON A GRINDER
150/HOUR	LOW	PACKS CHIPS IN THE BOTTOM OF A BLIND HOLE	MAY BE DONE BY HAND, ON A DRILL PRESS OR A LEAD SCREW TAPPING MACHINE
150/HOUR	LOW	MAY NOT FULLY FORM THE CREST OF THE THREAD	NO CHIPS ARE GENERATED, ALSO STRONGER THAN CUT THREADS
60/HOUR	MEDIUM	MUST BE RESET BY HAND	MAY BE RUN ON A LATHE, DRILL PRESS, OR VERTICAL MILL

Table 16 Welding

PROCESS	APPLICATION INFORMATION		
WELDING / TYPES	1. TYPICAL APPLICATIONS	2. PREPARATORY PROCESS	3. MATERIAL THICKNESS WELDABLE
1. SHIELDED METAL ARC	MANUAL WELDING INDOORS OR OUT	CLEAN AND GRIND	.50 INCH
2. SUBMERGED ARC	LONG WELDS IN CARBON AND LOW ALLOY STEEL	CLEAN, GRIND, AND APPLY FLUX	2.00 INCH PLATE
3. GAS METAL ARC (MIG)	WELDING OF ALL COMMERCIAL METALS	CLEAN	.50 INCH
4. GAS TUNGSTEN ARC (TIG)	HIGH QUALITY THIN PARTS	CLEAN	.25 INCH OR LESS
5. STUD ARC	ATTACHING STUD SCREWS, PINS, AND FASTENERS TO LARGER WORKPIECES	CLEAN	UNLIMITED
6. SPOT	SHEET METAL	CLEAN	.25 INCH
7. SEAM	CONTINUOUS SPOT WELDING, MAY BE OVERLAPPING	CLEAN	.125 INCH
8. PROJECTION	LOCALIZE SPOT WELDING	CLEAN	5 - 24 GAGE .025 - .218 INCH
9. BRAZING	JOINING DISSIMILAR METALS	CLEAN AND APPLY FLUX	UNLIMITED
10. SOLDERING	ELECTRONIC CONNECTIONS	CLEAN AND APPLY FLUX	THIN GAGE SHEET METAL AND WIRE
11. ELECTRON	ALL METALS THAT CAN BE ARC WELDED	CLEAN	UNLIMITED
12. ULTRASONIC	JOIN VERY LIGHT MATERIALS	CLEAN WELL AND DEOXIDIZE	FOIL TO WIRE OR WIRE TO WIRE

APPLICATION INFORMATION			
4. RELATIVE PRODUCTION VOLUME	5. RELATIVE TOOLING COST	6. DISADVANTAGE TO USE	7. COMMENTS
LOW	LOW	INTERRUPTIONS TO CHANGE STICK ELECTRODE AND DESLAGGING REQ'D	HIGHLY PORTABLE
LOW	LOW	MUST BE POSITIONED FLAT TO HOLD GRANULAR FLUX	WELD IS COMPLETED WITHOUT FLASH
HIGH	HIGH	EQUIPMENT COST AND PORTABILITY	SHOULD BE USED INDOORS
HIGH	MEDIUM	HIGH TEMPERATURES CREATE POOR WELD IN LOW MELDING TEMPERATURE METALS	HIGH LEVEL OF SKILL REQUIRED
HIGH	MEDIUM	NONE	STUD BECOMES ELECTRODE
HIGH	MEDIUM	NONE	MAY BE USED IN A SERIES
HIGH	MEDIUM	NONE	EMPLOYS CIRCULAR ELECTRODES
HIGH	MEDIUM	REQUIRES SPECIAL TOOLING	SMALL PROJECTION ON WORKPIECE LOCALIZES HEAT IN A SPECIFIC LOCATION
DEPENDS ON APPLICATION	LOW	NONE	BONDING MEDIUM HAS A MELTING POINT OVER 800 F
DEPENDS ON APPLICATION	LOW	NONE	FILLER HAS A MELTING POINT BELOW 800 F
LOW	VERY HIGH	MUST BE DONE IN A VACUUM	VACUUM CHAMBER LIMITS SIZE OF WORKPIECE
MEDIUM	MEDIUM	FINE PARTICLES OF METAL CAN BE THROWN OFF DURING PROCESS	HIGH SOUND FREQUENCIES EXCITE METAL SURFACE MOLECULES

APPENDIX IV

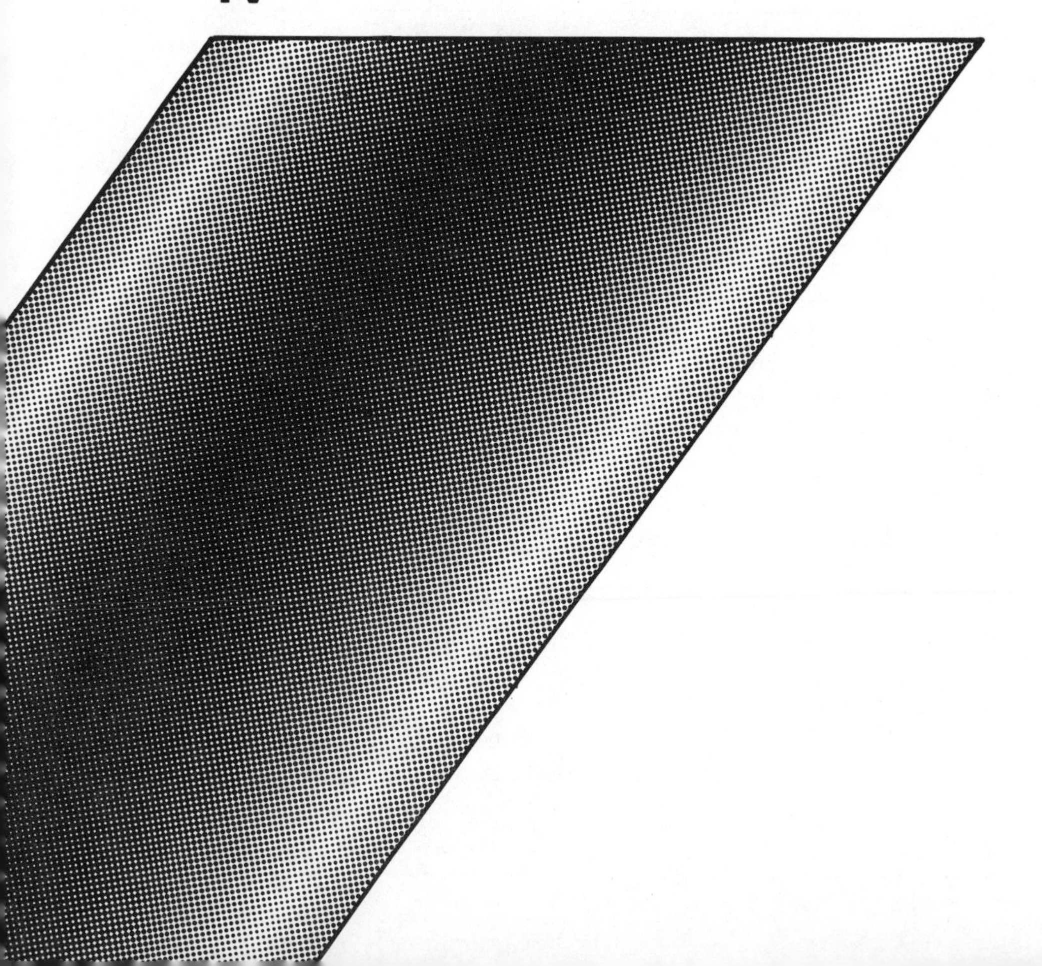

Machines

Table 1 Assembly
2 Bending
3 Deburring
4 Drilling
5 Forging
6 Furnaces
7 Gear Cutting
8 Grinding
9 Inspection
10 Material Handling
11 Milling
12 Plastics Processing
13 Saws
14 Turning

Table 1 Assembly

MACHINE	APPLICATION INFORMATION		
ASSEMBLY TYPES	1. TYPICAL APPLICATIONS	2. TYPICAL PROCESSES COMPLETED ON THIS MACHINE TYPE	3. WORK ENVELOPE OR MAXIMUM SIZE OF WORKPIECE
1. SINGLE STATION	PERFORMING A GIVEN FUNCTION MANY TIMES ON A SINGLE PART	ASSEMBLY, MACHINING, INSPECTION, MARKING, ETC.	UNLIMITED
2. SYNCHRONOUS DIAL	A NUMBER OF STATIONS ON A CIRCULAR INDEX TABLE BRINGS PARTS TO VARIOUS OPERATIONS	UNLIMITED	18 INCHES IN DIAMETER
3. SYNCHRONOUS IN-LINE	A SERIES OF FIXTURES MOVED IN A STRAIGHT LINE BY CHAIN OR BELT	UNLIMITED	COMPLETE CAR BODY
4. SYNCHRONOUS CARROUSEL	LIKE SYNCHRONOUS IN-LINE EXCEPT LINE FOLLOWS A RECTANGULAR PATH	UNLIMITED	COMPLETE CAR BODY
5. NON-SYNCHRONOUS	STATIONS OF MACHINE OPERATE INDEPENDENTLY	UNLIMITED	COMPLETE CAR BODY
6. CONTINUOUS	A WORK PERFORMED ON THE FLY (I.E. BOTTLING)	ASSEMBLY, FILLING, INSPECTION, MARKING, ETC.	COMPLETE CAR BODY
7. PARTS FEEDER (SELECTORS)	HOPPER, VIBRATORY BOWLS, AND MAGAZINES	FEEDING OF PARTS TO WORKSTATION	RELATIVELY SMALL COMPONENTS
8. PARTS FEEDER (ORIENTERS)	HOPPER, VIBRATORY BOWLS, AND MAGAZINES	ORIENTING THE PART SO IT CAN BE PROPERLY ASSEMBLED	RELATIVELY SMALL COMPONENTS
9. PICK AND PLACE	PICK A SMALL COMPONENT UP FROM ONE SPOT AND PLACE IT AT ANOTHER LOCATION	PART TRANSFER AND ASSEMBLY	50 LBS.
10. ROBOTS	MOVING PARTS IN LOW VOLUME OR IN SPECIAL ATMOSPHERIC APPLICATIONS	ASSEMBLY, WELDING, INSPECTION, GLUING, MARKING, ETC.	1,000 LBS.
11.			
12.			

APPLICATION INFORMATION			
4. PRODUCTION RATE	5. RELATIVE EQUIPMENT COST	6. FLOOR SPACE REQUIRED	7. DISADVANTAGE
1.000/HOUR	MEDIUM	VARIABLE	NORMALLY A SPECIAL MACHINE GOOD FOR ONLY ONE OPERATION
3.000/HOUR	MEDIUM	VARIABLE	NORMALLY A SPECIAL MACHINE GOOD FOR ONLY A SERIES OF OPERATIONS
1.500/HOUR	MEDIUM	VARIABLE	MAXIMUM NUMBER OF WORKSTATIONS TYPICALLY 20
1.500/HOUR	HIGH	VARIABLE	COST
1.200/HOUR	HIGH	VARIABLE	EXTRA PALLETS MUST BE PROVIDED FOR PARTS TO ACCUMULATE AT ONE STATION
400/MINUTE	HIGH	VARIABLE	COMPLEX SYSTEMS ARE SUBJECT TO MALFUNCTIONS
UNLIMITED	LOW	MINIMAL	THIS DEVICE IS ONLY A SMALL PART OF AN ASSEMBLY MACHINE
UNLIMITED	LOW	MINIMAL	THIS DEVICE IS ONLY A SMALL PART OF AN ASSEMBLY MACHINE
1.000/HOUR	LOW	VARIABLE (BASED ON SWING)	ONLY FOLLOWS ONE FIXED PATH
250/HOUR	MEDIUM	VARIABLE (BASED ON SWING)	RELATIVELY SLOW

Table 2 Bending

MACHINE	APPLICATION INFORMATION		
BENDING, FORMING SHEARING, AND NOTCHING (INCLUDES PRESS TYPES) TYPES	1. TYPICAL APPLICATIONS	2. TYPICAL PROCESSES COMPLETED ON THIS MACHINE TYPE	3. WORK ENVELOPE OR MAXIMUM SIZE OF WORKPIECE
1. BOX AND PAN BRAKE	BENDING UP OF SHEET METAL BOXES AND PANS	SHEET METAL BENDING	60 IN. WIDE SHEET STOCK
2. PRESS BRAKE	BENDING OF SHEET METAL	THE FULL RANGE OF SHEET METAL PROCESSES	24 FEET WIDE SHEET STOCK
3. ROLL FORMER	BENDING OF PARALLEL BENDS	BENDING OF PARALLEL BENDS	DEPENDS ON MACHINE (MACHINES ARE OFTEN SPECIALLY DESIGNED)
4. STRETCH FORMER	STRETCHING AND FORMING OF AIRCRAFT SKINS AND OTHER SIMILAR PARTS	STRETCHING AND FORMING OF SHEET METAL	DEPENDS ON MACHINE (MACHINES ARE OFTEN SPECIALLY DESIGNED)
5. METAL SPINNER	SPINNING OF CUP SHAPES FROM SHEET METAL	ASIDE FROM SPINNING, OTHER LATHE OPERATIONS	72 IN. DIAMETER
6. TUBE BENDER	BENDING OF TUBES FOR AUTOMOTIVE BRAKE LINES, GAS LINES, ETC.	TUBE BENDING	3 IN. TUBE DIA.
7. IRONWORKER	LOW VOLUME SHEET METAL FABRICATION	BLANKING, PIERCING, BENDING, SHEARING, NOTCHING, ETC.	12 IN. WIDE SHEET STOCK
8. N.C. TURRET PUNCH	PUNCHING HOLE PATTERNS IN SHEET METAL FOR THE FURNITURE INDUSTRY	PUNCHING AND NOTCHING	A 12 FOOT SQUARE
9. MECHANICAL PRESS	PROGRESSIVE DIE WORK	THE FULL RANGE OF SHEET METAL PROCESSES	4,000 TON PRESS WITH 360 IN. X 180 IN. BED
10. HYDRAULIC PRESS	PROGRESSIVE DIE WORK	THE FULL RANGE OF SHEET METAL PROCESSES	4,000 TON PRESS WITH 360 IN. X 180 IN. BED
11. OPEN BACK INCLINABLE	SIMPLE BLANKED AND STAMPED PARTS	THE FULL RANGE OF SHEET METAL PROCESSES	250 TONS WITH A 36 IN. X 36 IN. BED
12. GUILLOTINE SHEAR	SHEARING SHEET, PLATE, AND SMALL BAR STOCK	SHEARING	420 IN. WIDE SHEET STOCK

APPLICATION INFORMATION			
4. PRODUCTION RATE	5. RELATIVE EQUIPMENT COST	6. FLOOR SPACE REQUIRED	7. DISADVANTAGE
NOT A PRODUCTION MACHINE	VERY LOW	18 SQ. FT.	CANNOT BE AUTOMATED
240 BENDS/HOUR	LOW	UP TO 100 SQ. FT.	TYPICALLY A LOW VOLUME MACHINE
100-300 FEET/MINUTE	HIGH	VARIES WITH MACHINE DESIGN (200 SQ. FT. MIN.)	MUST HAVE OVER A MILLION FEET OF WORK PER YEAR TO JUSTIFY
60/HOUR	MEDIUM	VARIES GREATLY WITH APPLICATION	NOT SUITED FOR PROGRESSIVE OPERATION
PROCESS CONFINED TO RUNS OF 1,000 PIECES OR LESS (PRODUCTION RATE VARIES)	MEDIUM	UP TO 100 SQ. FT.	SOME LESS DUCTILE MATERIALS ARE NOT SUITABLE FOR METAL SPINNING
240 BENDS/HOUR	LOW	18 SQ. FT.	NONE
NOT A PRODUCTION MACHINE	MEDIUM	18 SQ. FT.	LIMITATIONS OF PART SIZE
DEPENDS ON THE NUMBER OF HOLES TO BE PUNCHED IN A SINGLE SHEET	MEDIUM	150 SQ. FT.	NONE
1,500 STROKES/MINUTE	HIGH	150 SQ. FT.	STROKE LENGTH FIXED
UP TO 900 STROKES/MINUTE	HIGH	150 SQ. FT.	SLOWER THAN MECHANICAL
250 STROKES/MINUTE	MEDIUM	24 SQ. FT.	LIMITED TONNAGES AND FRAME MAY SPRING
50-80 STROKES/MINUTE	MEDIUM	36 SQ. FT.	MAY REQUIRE A NUMBER OF PIECES OF SUPPORT EQUIPMENT

Table 3 Deburring

MACHINE	APPLICATION INFORMATION		
DEBURRING TYPES	1. TYPICAL APPLICATIONS	2. TYPICAL PROCESSES COMPLETED ON THIS MACHINE TYPE	3. WORK ENVELOPE OR MAXIMUM SIZE OF WORKPIECE
1. ABRASIVE BELT	EXTERNAL DIAMETERS AND FLAT DISKS	DEBURRING	VARIOUS
2. BARREL TUMBLER	DEBURRING OF FORGED, MACHINED AND STAMPED PARTS	DEBURRING POLISHING	A 12 INCH CUBE
3. ELECTRO-CHEMICAL	DEBURRING OF SMALL OR FINE PARTS	DEBURRING	SMALL TO LARGE SHEETS
4. EXPLOSIVE (THERMAL)	DEBURRING METAL PARTS	DEBURRING	A 12 INCH CUBE
5. HONING	INTERNAL DIAMETERS AND FLAT SURFACES	DEBURRING, SURFACE FINISHING	A 6 INCH CUBE
6. LAPPING	INTERNAL DIAMETERS AND FLAT SURFACES	DEBURRING, SURFACE FINISHING	A 6 INCH CUBE
7. PEDESTAL GRINDER	SNAG GRINDING AND WIRE BRUSHING ACCESSIBLE EDGES	SNAGGING, DEBURRING, BUFFING	UNLIMITED
8. SAND BLASTING	CHANGING COLOR, DEBURRING AND REMOVING OXIDATION	DEBURRING, SURFACE CLEANING	AN 18 INCH CUBE
9. SHOT PEENING	CLEANING, PEENING, WHILE INCREASING FATIGUE STRENGTH OF METAL PARTS	CLEANING AND PEENING	AN 18 INCH CUBE
10. SPIRAL VIBRATORY	CONTINUOUS DEBURRING AND POLISHING OF PARTS	CLEANING, DEBURRING, POLISHING	A 12 INCH CUBE
11. TUB VIBRATORY	DEBURRING OF FORGED, MACHINED AND STAMPED PARTS	DEBURRING POLISHING	A 12 INCH CUBE
12. VAPOR BLAST	DEBURRING OR CHANGING THE SURFACE FINISH OF METAL PARTS	ROUGHING UP A SURFACE	VARIOUS SPECIAL SHAPES CAN BE ACCOMMODATED

APPLICATION INFORMATION			
4. PRODUCTION RATE PER HOUR	5. RELATIVE EQUIPMENT COST	6. FLOOR SPACE REQUIRED	7. DISADVANTAGE
600	LOW	12 SQ. FT.	VENTILATION CONTROL REQUIRED
BATCH PROCESS 4 - 12 HOURS	MEDIUM	18 SQ. FT.	MUST BE FOLLOWED BY A CLEANING OPERATION
120	LOW	18 SQ. FT.	MUST BE FOLLOWED BY A CLEANING OPERATION
80	MEDIUM	18 SQ. FT.	NONE
30	HIGH	15 SQ. FT.	MUST BE FOLLOWED BY A CLEANING OPERATION
BATCH PROCESS 20 MINS	HIGH	36 SQ. FT.	MUST BE FOLLOWED BY A CLEANING OPERATION
240	LOW	4 SQ. FT.	DIFFICULT TO AUTOMATE
60	LOW	15 SQ. FT.	MUST BE FOLLOWED BY A CLEANING OPERATION
60	HIGH	36 SQ. FT.	REQUIRES SPECIAL PART FIXTURING
60	HIGH	36 SQ. FT.	NONE
BATCH PROCESS 1 HOUR	MEDIUM	24 SQ. FT.	BURRS CONTAMINATES MEDIA
60	MEDIUM	15 SQ. FT.	REQUIRES SPECIAL PART SET UP

Table 4 Drilling

MACHINE	APPLICATION INFORMATION		
DRILLING TYPES	1. TYPICAL APPLICATIONS	2. TYPICAL PROCESSES COMPLETED ON THIS MACHINE TYPE	3. WORK ENVELOPE OR MAXIMUM SIZE OF WORKPIECE
1. SINGLE SPINDLE SENSITIVE-HAND FEED	DRILLING ONE HOLE AT A TIME IN SMALL PARTS	DRILLING, REAMING, COUNTERBORING, SPOTFACING, TAPPING	A 12 INCH CUBE
2. TURRET TYPE BENCH MOUNTED HAND FEED	PERFORMING A VARIETY OF OPERATIONS WITHOUT CHANGING TOOLS	DRILLING, REAMING, COUNTERBORING, SPOTFACING, TAPPING	A 6 INCH CUBE
3. SINGLE SPINDLE POWER FEED	ACCURATELY DRILLING ONE HOLE AT A TIME IN LARGER PARTS	DRILLING, REAMING, COUNTERBORING, SPOTFACING, TAPPING	AN 18 INCH CUBE
4. SINGLE SPINDLE POWER FEED WITH MULTIPLE SPINDLE HEAD	ACCURATELY DRILLING A NUMBER OF HOLES SIMULTANEOUSLY	DRILLING, REAMING, COUNTERBORING, SPOTFACING, TAPPING	AN 18 INCH CUBE
5. GANG DRILL	HIGH PRODUCTION MULTIPLE HOLE OR MULTIPLE OPERATION PROCESSING	DRILLING, REAMING, COUNTERBORING, SPOTFACING, TAPPING	A 12 INCH CUBE/DRILL
6. MULTIPLE SPINDLE WITH FIXED CENTER HEADS	VERY HIGH PRODUCTION, FIXED PATTERN MULTIPLE HOLE DRILLING	DRILLING, REAMING, COUNTERBORING, SPOTFACING, TAPPING	A 24 INCH CUBE
7. TURRET TYPE WITH POWER FEED	PERFORMING A VARIETY OF OPERATIONS WITHOUT CHANGING TOOLS	DRILLING, REAMING, COUNTERBORING, SPOTFACING, TAPPING	AN 18 INCH CUBE
8. RADIAL DRILL	DRILLING HOLES IN LARGE, IRREGULARLY SHAPED PARTS	DRILLING, REAMING, COUNTERBORING, SPOTFACING, TAPPING	A 36 INCH CUBE
9. HORIZONTAL FIXED TABLE ADVANCING SPINDLE	DRILLING EXTREMELY DEEP HOLES	DRILLING, REAMING, COUNTERBORING, SPOTFACING, TAPPING	12 X 12 X 60 INCHES
10. AUTOMATIC SELF FEEDING DRILL UNIT	PORTABLE DRILLING AND SPECIAL DRILLING MACHINE CONSTRUCTION	DRILLING, REAMING, COUNTERBORING, SPOTFACING, TAPPING	UNLIMITED
11. BED TYPE FIXED SPINDLE (DIGITAL READOUT)	DRILLING A SERIES OF HOLES IN A LARGE WORKPIECE	DRILLING, REAMING, COUNTERBORING, SPOTFACING, TAPPING	A 12 INCH CUBE
12. SLIDING HEAD DRILL, FIXED SPINDLE (NC)	AUTOMATIC SERIES OF OPERATIONS WITHOUT CHANGING TOOLS	DRILLING, REAMING, COUNTERBORING, SPOTFACING, TAPPING	A 12 INCH CUBE

APPLICATION INFORMATION			
4. PRODUCTION RATE	5. RELATIVE EQUIPMENT COST	6. FLOOR SPACE REQUIRED (DOES NOT INCLUDE OPERATOR)	7. DISADVANTAGE
120/HOUR	LOW	9 SQ. FT.	NOT A PRODUCTION MACHINE
20/HOUR	LOW	NONE	INDEXING AND MULTIPLE TOOLS MAKE THIS A SLOW MACHINE
80/HOUR	MEDIUM	9 SQ. FT.	MULTIPLE SPINDLE HEAD REQUIRED TO INCREASE PRODUCTION CAPABILITIES
80/HOUR	MEDIUM	9 SQ. FT.	NONE
30/HOUR	HIGH	24 SQ. FT.	NONE
80/HOUR	HIGH	12 SQ. FT.	ONLY GOOD FOR A SINGLE HOLE PATTERN
20/HOUR	MEDIUM	15 SQ. FT.	INDEXING AND MULTIPLE TOOLS MAKE THIS A SLOW MACHINE
12/HOUR	LOW	24 SQ. FT.	SLOW AND CUMBERSOME
12/HOUR	HIGH	20 SQ. FT.	ONLY GOOD FOR DEEP HOLE DRILLING
80/HOUR IF FIXED IN A SPECIAL MACHINE	VARIABLE	VARIABLE	NONE
6/HOUR	LOW	32 SQ. FT.	NONE
6/HOUR	LOW	60 SQ. FT.	NONE

Table 5 Forging

MACHINE	APPLICATION INFORMATION		
FORGING TYPES	1. TYPICAL APPLICATIONS	2. TYPICAL PROCESSES COMPLETED ON THIS MACHINE TYPE	3. SIZE OF MACHINE
1. GRAVITY DROP HAMMER	WHERE CONSTANT BLOWS ARE REQUIRED	OPEN DIE FORGING	500 - 16,000 LBS.
2. DIE FORGER HAMMER	WHERE CONSTANT BLOWS ARE REQUIRED	OPEN DIE FORGING	1,000 - 70,000 LBS.
3. COUNTERBLOW HAMMER	WHERE CONSTANT BLOWS ARE REQUIRED	OPEN DIE FORGING	4,000 - 66,000 FT. LBS.
4. OPEN DIE FORGING HAMMER	WHERE CONSTANT BLOWS ARE REQUIRED	OPEN DIE FORGING	8,000 - 220,000 LBS.
5. MECHANICAL PRESS	ACCURATE AND CLOSE TOLERANCE PARTS	CLOSED DIE (SQUEEZING) FORGING	3,000 - 16,000 TONS
6. HYDRAULIC PRESS	EXTRUSION TYPE FORGING	CLOSED DIE (SQUEEZING) FORGING	700 - 14,000 TONS
7. SCREW PRESS	FORGING NONSYMMETRICAL PARTS	OPEN AND CLOSED DIE FORGING	160 - 4,000 TONS
8. HOT FORMER	HIGH PRODUCTION OF CLOSE TOLERANCE PARTS	FEEDING, SHEARING AND FORGING	CAN SHEAR BARS UP TO 3.5 DIA. X 5.88 LG.
9. FORGING ROLLS	PREFORMING BLANKS, CRANKSHAFTS, AND AXLES	FORGING AND DESCALING	ROLLING BLANKS UP TO 5 INCHES IN DIAMETER
10. WEDGE ROLLING	PREFORMING OF BALLS, TAPERS, AND UNDERCUTS	ROLLING SHAPES FROM BAR STOCK	ROLLING BLANKS UP TO 5 INCHES IN DIAMETER
11. RADIAL FORGING	PRECISION FORMING BARS AND TUBES	REPEATED HIGH IMPACT BLOWS	60 - 250 TONS
12. RING ROLLING	SEAMLESS RINGS	REDUCING THE WALL THICKNESS OF DONUT SHAPES	ROLLING RINGS UP TO 24 INCHES IN DIAMETER

APPLICATION INFORMATION			
4. PRODUCTION RATE	5. RELATIVE EQUIPMENT COST	6. FLOOR SPACE REQUIRED	7. DISADVANTAGE
45 - 60 BLOWS PER MINUTE	LOW	100 SQ. FT.	DROP HEIGHT NOT ADJUSTABLE
60 - 100 BLOWS PER MINUTE	LOW	100 SQ. FT.	DROP HEIGHT NOT ADJUSTABLE
100 - 170 BLOWS PER MINUTE	LOW	100 SQ. FT.	WORKPIECE JUMPS FROM DIE. DIE MISMATCH
60 BLOWS PER MINUTE	LOW	100 SQ. FT.	NOT FOR PRELIMINARY FORGING OPERATIONS
30 - 100 BLOWS PER MINUTE	HIGH	100 SQ. FT.	RELATIVELY SLOW
RAM SPEED 25 - 300 INCHES PER MINUTE	HIGH	100 SQ. FT.	FULL FORCE TAKES PLACE ONLY NEAR CENTER OF BED
12 - 50 BLOWS PER MINUTE	HIGH	100 SQ. FT.	NONE
45 - 180 BLOWS PER MINUTE	HIGH	500 SQ. FT.	NONE
———	MEDIUM	100 SQ. FT.	NONE
600 - 1200/HOUR	MEDIUM	100 SQ. FT.	NONE
600 - 1200/HOUR	MEDIUM	100 SQ. FT.	NONE
600 - 1200/HOUR	MEDIUM	100 SQ. FT.	NONE

Table 6 Furnaces

MACHINE	APPLICATION INFORMATION		
FURNACES TYPES	1. TYPICAL APPLICATIONS	2. TYPICAL PROCESSES COMPLETED ON THIS MACHINE TYPE	3. WORK ENVELOPE OR MAXIMUM SIZE OF WORKPIECE
1. BOX (BATCH)	LABORATORY OPERATIONS	HEAT TREATING OF SINGLE PARTS OR SMALL LOTS	A SMALL BASKET
2. CAR BOTTOM (BATCH)	HEAT TREATING LARGE CASTINGS AND FORGINGS	VARIOUS HEAT TREATING PROCESSES	A LARGE BASKET
3. ELEVATOR (BATCH)	HEAT TREATING LARGE CASTINGS AND FORGINGS	VARIOUS HEAT TREATING PROCESSES	A LARGE BASKET
4. BELL TYPE (BATCH)	HEAT TREATING WHERE PROTECTIVE ATMOSPHERE IS REQUIRED	VARIOUS HEAT TREATING PROCESSES	A LARGE BASKET
5. PIT (POT) (BATCH)	HEAT TREATING LARGE CASTINGS AND FORGINGS	VARIOUS HEAT TREATING PROCESSES	A LARGE BASKET
6. ROTARY-HEARTH (CONTINUOUS)	HEAT TREATING OF PARTS THAT HAVE TO BE INDIVIDUALLY HANDLED	VARIOUS HEAT TREATING PROCESSES	BASKETS OR TRAYS
7. STRAIGHT CHAMBER (CONTINUOUS)	HEAT TREATING OF PARTS IN BASKETS	VARIOUS HEAT TREATING PROCESSES	BASKETS OR TRAYS
8. WITH INTERNAL QUENCH TANK (CONTINUOUS)	HEAT TREATING AND QUENCHING WITHIN A CONTROLLED ATMOSPHERE	VARIOUS HEAT TREATING PROCESSES	BASKETS OR TRAYS
9. SALT BATH	A WIDE VARIETY OF HEAT TREATING PROCESSES REQUIRING RAPID HEATING OF THE WORKPIECE	VARIOUS HEAT TREATING PROCESSES	BASKETS OR TRAYS
10. VACUUM	HIGH TEMPERATURE PROCESSING OF REACTIVE AND REFACTORY METALS	STRESS RELIEVING, ANNEALING, HARDENING, TEMPERING, AND SOLUTION TREATING	BASKETS OR TRAYS
11. FLUIDIZED BED	A WIDE VARIETY OF HEAT TREATING PROCESSES REQUIRING RAPID HEATING OF THE WORKPIECE	VARIOUS HEAT TREATING PROCESSES	BASKETS OR TRAYS
12.			

APPLICATION INFORMATION			
4. PRODUCTION RATE	5. RELATIVE EQUIPMENT COST	6. FLOOR SPACE REQUIRED	7. DISADVANTAGE
NOT A PRODUCTION FURNACE	LOW	24 SQ. FT.	NONE
DEPENDS UPON HEAT TREATMENT PROCESS	MEDIUM	64 SQ. FT.	NONE
DEPENDS UPON HEAT TREATMENT PROCESS	MEDIUM	64 SQ. FT.	NONE
DEPENDS UPON HEAT TREATMENT PROCESS	MEDIUM	64 SQ. FT.	SLOW LOAD AND UNLOAD
DEPENDS UPON HEAT TREATMENT PROCESS	MEDIUM	100 SQ. FT. OR MORE	SLOW LOAD AND UNLOAD
DEPENDS UPON HEAT TREATMENT PROCESS	HIGH	100 SQ. FT. OR MORE	RATE OF TRAVEL THROUGH FURNACE MUST BE COORDINATED
DEPENDS UPON HEAT TREATMENT PROCESS	HIGH	500 SQ. FT.	RATE OF TRAVEL THROUGH FURNACE MUST BE COORDINATED
DEPENDS UPON HEAT TREATMENT PROCESS	HIGH	500 SQ. FT.	RATE OF TRAVEL THROUGH FURNACE MUST BE COORDINATED
DEPENDS UPON HEAT TREATMENT PROCESS	LOW	200 SQ. FT.	BUOYANT WORKPIECES HARD TO HANDLE
LIMITED BY SIZE OF VACUUM CHAMBER	HIGH	500 SQ. FT.	COST
DEPENDS UPON HEAT TREATMENT PROCESS	MEDIUM	200 SQ. FT.	NONE

Table 7 Gear Cutting

MACHINE	APPLICATION INFORMATION		
GEAR CUTTING TYPES	1. TYPICAL APPLICATIONS	2. TYPICAL PROCESSES COMPLETED ON THIS MACHINE TYPE	3. WORK ENVELOPE OR MAXIMUM SIZE OF WORKPIECE
1. BROACHING	INTERNAL AND EXTERNAL SPUR TEETH	BROACHING	10 IN. DIA. 4 IN. FACE WIDTH
2. FACE MILLING	CUTTING A VARIETY OF BEVEL AND HYPOID GEARS	GENERATING GEAR TEETH	50 IN. DIA. 10 IN. FACE WIDTH
3. G TRAC	SPUR AND HELICAL GEARS	ROUGHING AND FINISHING SPUR AND HELICAL GEARS	14 IN. DIA. 7.5 IN. STACK
4. GRINDING	AIRCRAFT AND HIGH PITCH LINE VELOCITY GEARS	GEAR FINISHING	EQUAL TO THAT OF THE MAXIMUM ROUGHING MACHINE CAPABILITIES
5. HOBBING	GENERATING EXTERNAL SPUR AND HELICAL GEARS	HOBBING	24 IN. DIA. 24 IN. FACE WIDTH
6. LAPPING	FINISHING OF HARDENED GEARS	GEAR FINISHING	EQUAL TO THAT OF THE MAXIMUM ROUGHING MACHINE CAPABILITIES
7. REVACYCLE	CUTTING LOW SPEED DIFFERENTIAL PINION AND SIDE BEVEL GEARS	ROUGHING AND FINISHING STRAIGHT BEVEL GEARS	6 IN. DIA. 1.5 IN. FACE WIDTH
8. PLANER	ROUGHING LARGE BEVEL GEARS	PLANING	UNLIMITED
9. ROLLING	FORMING OF SPLINES ON AXLES AND OTHER SHAFTS	GEAR FINISHING	EQUAL TO THAT OF THE MAXIMUM ROUGHING MACHINE CAPABILITIES
10. SHAPING	GENERATING INTERNAL AND EXTERNAL GEAR TEETH	GEAR FINISHING	50 IN. DIA. 9.84 IN. FACE WIDTH
11. SHAVING	FINISHING OF SOFT GEARS	GEAR FINISHING	EQUAL TO THAT OF THE MAXIMUM ROUGHING MACHINE CAPABILITIES
12. SHEAR SPEED	CUTTING STRAIGHT SPUR TEETH	A TYPE OF POT BROACHING	20 IN. DIA. 6 IN. FACE WIDTH

APPLICATION INFORMATION			
4. PRODUCTION RATE	5. RELATIVE EQUIPMENT COST	6. FLOOR SPACE REQUIRED	7. DISADVANTAGE
150/HOUR	MEDIUM	36 SQ. FT.	BROACH STROKE LIMITS AMOUNT OF STOCK THAT CAN BE REMOVED
40/HOUR	MEDIUM	64 SQ. FT.	TAKES TWO MACHINES. ONE ROUGH, ONE FINISH
180-600/HOUR	HIGH	80 SQ. FT.	NONE
6/HOUR	HIGH	64 SQ. FT.	EXPENSIVE SECONDARY DEDICATED MACHINE
75/HOUR	MEDIUM	16 SQ. FT.	NONE
60/HOUR	LOW	16 SQ. FT.	EXPENSIVE SECONDARY DEDICATED MACHINE
60/HOUR	HIGH	64 SQ. FT.	LIMITED TO STRAIGHT TOOTH BEVEL GEARS
20/HOUR	MEDIUM	64 SQ. FT.	TAKES TWO MACHINES. ONE ROUGH, ONE FINISH
500-1200/HOUR	LOW	16 SQ. FT.	EXPENSIVE SECONDARY DEDICATED MACHINE
50/HOUR	MEDIUM	64 SQ. FT.	EXPENSIVE SECONDARY DEDICATED MACHINE
60/HOUR	LOW	16 SQ. FT.	EXPENSIVE SECONDARY DEDICATED MACHINE
240/HOUR	HIGH	64 SQ. FT.	LIMITED TO SPUR TEETH IN A SMALL RANGE OF DIAMETERS

Table 8 Grinding

MACHINE	APPLICATION INFORMATION		
GRINDING TYPES	1. TYPICAL APPLICATIONS	2. TYPICAL PROCESSES COMPLETED ON THIS MACHINE TYPE	3. WORK ENVELOPE OR MAXIMUM SIZE OF WORKPIECE
1. CYLINDRICAL (PLAIN)	GRINDING AXLES, SPINDLES, SHAFTS, ETC.	SAME AS MACHINING TYPE	36 INCH DIA. X 192 INCHES LONG
2. CYLINDRICAL (UNIVERSAL)	GRINDING SHAFTS, SHOULDERS AND FACES	SAME AS MACHINING TYPE	9 INCH DIA. X 72 INCHES LONG
3. CYLINDRICAL (THREAD AND FORM)	PLUNGE GRINDING OF SPECIAL FORMS	SAME AS MACHINING TYPE	24 INCH DIA. X 96 INCHES LONG
4. INTERNAL (CHUCKING)	GRINDING INTERNAL DIAMETERS AND FACES	SAME AS MACHINING TYPE	.040 - 20.0 INCH DIA. 12 INCHES DEEP WITH THE 20 IN. DIA.
5. CENTERLESS	GRINDING OUTSIDE DIAMETERS WITHOUT CENTERS	SAME AS MACHINING TYPE	.020 - 12.0 INCH DIA. X 20 FEET LONG
6. CENTERLESS (ABRASIVE BELT)	THROUGH FEEDING OF ROUGH WORK, GRINDING AND DEBURRING	SAME AS MACHINING TYPE	9 INCH DIA. X 20 FEET LONG
7. CENTERLESS (END FEED)	GRINDING OUTSIDE DIAMETERS OF TAPERED AND SHOULDERED PARTS	SAME AS MACHINING TYPE	9 INCH DIA. X 36 INCHES LONG
8. CENTERLESS (INTERNAL)	GRINDING INSIDE DIAMETERS OF PARTS WITHOUT CENTERING OR CHUCKING	SAME AS MACHINING TYPE	.50 INCH DIA. AND UP X 1.625 - 10 INCH DEEP
9. SURFACE (HORIZONTAL)	GRINDING FLAT SURFACES	SAME AS MACHINING TYPE	12 INCHES WIDE X 36 INCHES LONG
10. SURFACE (ROTARY)	GRINDING FLAT SURFACES	SAME AS MACHINING TYPE	TABLE UP TO 96 INCHES IN DIAMETER
11. CAM	INTERNAL AND EXTERNAL CYLINDRICAL CAM TYPE SHAPES	SAME AS MACHINING TYPE	36 IN. DIAMETER
12. GEAR	EXTERNAL GEAR TEETH	SAME AS MACHINING TYPE	36 IN. DIAMETER

APPLICATION INFORMATION			
4. PRODUCTION RATE	5. RELATIVE EQUIPMENT COST	6. FLOOR SPACE REQUIRED	7. TYPICAL STOCK REMOVAL PER SIDE IN INCHES
60/HOUR	MEDIUM	VARIES WITH MACHINE SIZE	ROUGH: .002-.020 SEMIFINISH: .001-.010 FINISH: .0005-.005
60/HOUR	MEDIUM	VARIES WITH MACHINE SIZE	ROUGH: .002-.020 SEMIFINISH: .001-.010 FINISH: .0005-.005
60/HOUR	MEDIUM	VARIES WITH MACHINE SIZE	ROUGH: .002-.020 SEMIFINISH: .001-.010 FINISH: .0005-.005
60/HOUR	MEDIUM	VARIES WITH MACHINE SIZE	ROUGH: .002-.020 SEMIFINISH: .001-.010 FINISH: .0005-.005
600/HOUR	HIGH	VARIES WITH MACHINE SIZE	ROUGH: .002-.020 SEMIFINISH: .001-.010 FINISH: .0005-.005
1000/HOUR	MEDIUM	VARIES WITH MACHINE SIZE	ROUGH: .002-.020 SEMIFINISH: .001-.010 FINISH: .0005-.005
150/HOUR	HIGH	VARIES WITH MACHINE SIZE	ROUGH: .002-.020 SEMIFINISH: .001-.010 FINISH: .0005-.005
150/HOUR	HIGH	VARIES WITH MACHINE SIZE	ROUGH: .002-.020 SEMIFINISH: .001-.010 FINISH: .0005-.005
20/HOUR	LOW	VARIES WITH MACHINE SIZE	ROUGH: .002-.020 SEMIFINISH: .001-.010 FINISH: .0005-.005
200/HOUR BOTH SIDES	MEDIUM	VARIES WITH MACHINE SIZE	ROUGH: .002-.020 SEMIFINISH: .001-.010 FINISH: .0005-.005
60/HOUR	HIGH	VARIES WITH MACHINE SIZE	ROUGH: .002-.020 SEMIFINISH: .001-.010 FINISH: .0005-.005
10/HOUR	HIGH	VARIES WITH MACHINE SIZE	ROUGH: .002-.020 SEMIFINISH: .001-.010 FINISH: .0005-.005

Table 9 Inspection

MACHINE	APPLICATION INFORMATION		
INSPECTION TYPES	1. TYPICAL APPLICATIONS	2. TYPICAL PROCESSES COMPLETED ON THIS MACHINE TYPE	3. WORK ENVELOPE OR MAXIMUM SIZE OF WORKPIECE
1. BENCH CENTERS	INSPECTING SHAFTS AND DIAMETERS	CHECKING CIRCULAR AND TOTAL RUNOUT	12 IN. SWING
2. ROTATING SPINDLE ROUNDNESS MACHINE	CHECKING ROUNDNESS OF LONG SHAFTS	ROUNDNESS AND CONCENTRICITY MEASUREMENT	12 IN. DIA. X 48 IN. LONG
3. ENGINEERING MICROSCOPES	MAGNIFICATION	VIEWING DEFECTS NOT NORMALLY SEEN WITH THE NAKED EYE	VARIOUS
4. SURFACE ROUGHNESS MEASURING MACHINE	SURFACE ROUGHNESS INSPECTION	SURFACE ROUGHNESS INSPECTION	VARIOUS
5. GEAR TOOTH LEAD MEASURING MACHINE	HELICAL GEARS	INSPECTION OF THE THEORETICAL LEAD OF A HELICAL TOOTH	12 IN. PITCH DIA.
6. GEAR ROLLING MACHINE	CHECKING THE CENTER DISTANCE OF TWO MATING GEARS	CENTER DISTANCE ERROR MAY INDICATE ONE OR MORE PROBLEMS	2 - 9 IN. CENTER DISTANCE
7. GEAR SOUND TESTING MACHINE	CHECKING THE LEVEL OF NOISE OF TWO MATING GEARS	CHECKS NOISE UNDER VARIABLE SPEED AND GEAR LOADING	MACHINE DESIGNED TO SUIT
8. OPTICAL COMPARATOR	PROFILE VIEWING	INSPECTING THE PROFILE OF PRODUCTION PARTS AND CUTTING TOOLS	12 X 24 INCHES
9. COORDINATE COMPARATOR	INSPECTING TOOLS AND GAGES	ALL MANNER OF DIMENSIONAL MEASUREMENT	60 INCHES SQ. X 24 IN. HIGH
10. VISION	CHECKING ABSENCE OR PRESENCE OF PART FEATURES	SENSING	18 INCHES SQ.
11.			
12.			

APPLICATION INFORMATION			
4. PRODUCTION RATE	5. RELATIVE EQUIPMENT COST	6. FLOOR SPACE REQUIRED	7. DISADVANTAGE
30/HOUR ACTUAL PARTS INSPECTED	LOW	NONE	ONLY GOOD FOR PARTS WITH CENTERS OR BORES
30/HOUR ACTUAL PARTS INSPECTED	HIGH	12 SQ. FT.	LIMITED TO CHECKING ROUNDNESS AND CONCENTRICITY
NOT A PRODUCTION MACHINE	MEDIUM	NONE	REQUIRES SKILLED LABOR
60/HOUR ACTUAL PARTS INSPECTED	LOW	NONE	NONE
30/HOUR ACTUAL PARTS INSPECTED	HIGH	16 SQ. FT.	LIMITED TO GEAR INSPECTION
30/HOUR ACTUAL PARTS INSPECTED	HIGH	16 SQ. FT.	LIMITED TO GEAR INSPECTION
30/HOUR ACTUAL PARTS INSPECTED	HIGH	12 SQ. FT.	LIMITED TO GEAR INSPECTION
30/HOUR ACTUAL PARTS INSPECTED	MEDIUM	9 SQ. FT.	PART SIZE LIMITED
NOT A PRODUCTION MACHINE	HIGH	25 SQ. FT.	REQUIRES SKILLED LABOR
500/HOUR PASS THROUGH	HIGH	9 SQ. FT.	STILL BEING PERFECTED

Table 10 Material Handling

MACHINE	APPLICATION INFORMATION		
MATERIAL HANDLING / TYPES	1. TYPICAL APPLICATIONS	2. TYPICAL PROCESSES COMPLETED ON THIS MACHINE TYPE	3. WORK ENVELOPE OR MAXIMUM SIZE OF WORKPIECE
1. CHUTES	LINKING OTHER CONVEYORS OR TO PROVIDE ROOM FOR PART ACCUMULATION	TRANSFERRING PARTS OR PACKAGES	A 4 FOOT CUBE
2. CONVEYORS, BELTS, AND ROLLERS	TRANSFERRING MEDIUM WEIGHT LOADS OVER FIXED PATHS	MOVING BOXES	A 4 FOOT CUBE
3. OVERHEAD MONORAILS	MOVING INDIVIDUAL PARTS, RACKS OF PARTS, OR BASKETS OF PARTS	CLEANING, COATING PLATING, PAINTING, AND TRANSFERRING OF PARTS	THE SIZE OF AN AUTOMOBILE
4. SCREW AUGERS	HELIX REVOLVES TO PUSH LOOSE MATERIAL	MOVING GRAIN, CHEMICALS, CHIPS, ETC.	3 FEET IN DIAMETER
5. HOISTS	LIFTING HEAVY OBJECTS IN A CONFINED AREA	INTERMITTENT MOVES OF AWKWARD OBJECTS	UNLIMITED
6. CRANES	LIFTING HEAVY OBJECTS OVER A LARGE AREA	INTERMITTENT MOVES OF AWKWARD OBJECTS	UNLIMITED
7. HAND TRUCKS	MOVING UNIT LOADS IN SMALL CONFINED AREAS	MOVING PALLETS, TUBS, BASKETS, PANS, BARRELS, ETC.	TYPICALLY ONE PALLET AT A TIME
8. FORK LIFT TRUCKS	MOVING UNIT LOADS OVER DISTANCES OF 500 FEET	MOVING PALLETS, TUBS, BASKETS, PANS, BARRELS, ETC.	4 PALLETS HIGH
9. AUTOMATED GUILDED VEHICLE	MOVING LARGE, OFTEN FLAT UNIT LOADS OVER FIXED PATHS	MOVING LARGE, AWKWARD, AND LOOSE OBJECTS	UNLIMITED
10. AUTOMATED STORAGE AND RETRIEVAL SYSTEMS	STORING AND RETRIEVING A VARIETY OF UNIT LOADS	STORING AND RETRIEVING OF INDIVIDUAL ITEMS/LOADS	TYPICAL UNIT LOAD
11. PICK AND PLACE UNITS	TRANSFERRING OF INDIVIDUAL PARTS BETWEEN TWO MACHINES OR CONVEYORS	TRANSFERRING, ASSEMBLY, STAKING, MARKING, ETC.	DEPENDS UPON MODEL, RESTRICTED BY SIZE AND WEIGHT
12. ROBOTS	VARIED HANDLING OPERATIONS IN A VARIETY OF ENVIRONMENTS	MOVING PARTS, ASSEMBLY, WELDING, INSPECTION, ETC.	DEPENDS UPON MODEL, RESTRICTED BY WEIGHT, NORMALLY UNDER 1,000 LBS.

APPLICATION INFORMATION			
4. PRODUCTION RATE	5. RELATIVE EQUIPMENT COST	6. FLOOR SPACE REQUIRED	7. DISADVANTAGE
720 PARTS/HOUR	LOW	VARIES WITH APPLICATION	PARTS MAY BE DAMAGED. PACKAGES MAY JAM
1,800 BOXES/HOUR	MEDIUM	VARIES WITH APPLICATION	LARGE WORK-IN-PROCESS INVENTORY
1 FOOT/SEC. (MAXIMUM MONORAIL TRAVEL SPEED)	MEDIUM	(NONE) CEILING SUSPENDED	FIXED PATH NOT FLEXIBLE
VARIES WITH MATERIAL BEING MOVED	MEDIUM	(NONE) BELOW THE FLOOR	TOUGH TO CLEAN
NOT APPLICABLE	MEDIUM	(NONE) CEILING SUSPENDED	SLOW AND MAY BE DANGEROUS
NOT APPLICABLE	MEDIUM	4 SQ. FT. PLUS THE PATH OF TRAVEL	SLOW AND MAY BE DANGEROUS
NOT APPLICABLE	MEDIUM	15 SQ. FT.	REQUIRES A MATERIAL HANDLER (I.E. LABOR)
NOT APPLICABLE	MEDIUM	24 SQ. FT.	REQUIRES A MATERIAL HANDLER (I.E. LABOR)
NOT APPLICABLE	HIGH	24 SQ. FT. PLUS THE PATH OF TRAVEL	NONE
UP TO 100 PICKS/HOUR	HIGH	5,000 SQ. FT.	LARGE WORK-IN-PROCESS INVENTORY
UP TO 1000 /HOUR	MEDIUM	16 SQ. FT.	CONSIDERED HARD AUTOMATION (I.E. MOVEMENT AND PATH ARE FIXED)
100-1,000 CYCLES/HOUR	HIGH	100 SQ. FT.	MUST BE PROGRAMMED. MOVEMENT DISTANCE LIMITED

Table 11 Milling

MACHINE	APPLICATION INFORMATION		
MILLING TYPES	1. TYPICAL APPLICATIONS	2. TYPICAL PROCESSES COMPLETED ON THIS MACHINE TYPE	3. WORK ENVELOPE OR MAXIMUM SIZE OF WORKPIECE
1. PLAIN	SLAB AND GANG MILLING	FACE MILLING, SHELL END MILLING	A 12 INCH CUBE
2. UNIVERSAL	GEARS, MILLING CUTTERS, DRILLS END MILLS	MILLING HELIXES	A 12 INCH CUBE
3. VERTICAL	TOOL ROOM AND LIGHT PRODUCTION	FACE AND END MILLING	A 12 INCH CUBE
4. RAM HEAD	TOOL ROOM AND LIGHT PRODUCTION	FACE AND END MILLING	A 12 INCH CUBE
5. TURRET RAM	TOOL ROOM	FACE AND END MILLING	A 12 INCH CUBE
6. HORIZONTAL	PRODUCTION	SLAB AND GANG MILLING	AN 18 INCH CUBE
7. DUPLEX	PRODUCTION MILLING OF OPPOSING PARALLEL SURFACES	FACE MILLING	AN 18 X 18 X 48 INCH RECTANGLE
8. VERTICAL FIXED BED	HEAVY PRODUCTION MILLING	FACE AND END MILLING	AN 18 X 18 X 48 INCH RECTANGLE
9. TRACER	COMPLEX SHAPES, DIESINKING	END AND BALL END MILLING	AN 18 X 18 X 48 INCH RECTANGLE
10. NC/CNC	COMPLEX PART WITH MULTIPLE OPERATIONS, ALSO CALLED MACHINING CENTERS	DRILL, MILL, BORE, REAM, COUNTERBORE, HOLLOW MILL, TAP, TREPAN	AN 18 INCH CUBE
11. FIXED BRIDGE	MILLING LONG SLOTS AND FEATURES	FACE AND END MILLING	AN 18 X 18 X 48 INCH RECTANGLE
12. MOVING BRIDGE GANTRY TYPE	MILLING LARGE SURFACE AREAS	FACE AND END MILLING	AN 48 X 48 X 18 INCH THICK RECTANGLE

APPLICATION INFORMATION			
4. PRODUCTION RATE	5. RELATIVE EQUIPMENT COST	6. FLOOR SPACE REQUIRED	7. TYPES OF MOVEMENT
SEE PROCESS CHART ON MILLING	LOW	20 SQ. FT.	KNEE, SADDLE, AND TABLE
SEE PROCESS CHART ON MILLING	LOW	20 SQ. FT.	KNEE, SADDLE TABLE, AND TABLE SWIVEL
SEE PROCESS CHART ON MILLING	LOW	20 SQ. FT.	KNEE, SADDLE TABLE, AND VERTICAL SPINDLE
SEE PROCESS CHART ON MILLING	MEDIUM	24 SQ. FT.	KNEE, SADDLE TABLE, VERTICAL SPINDLE AND RAM
SEE PROCESS CHART ON MILLING	MEDIUM	20 SQ. FT.	KNEE, SADDLE TABLE, RAM, AND RAM ON A ROTATING TURRET
SEE PROCESS CHART ON MILLING	MEDIUM	32 SQ. FT.	TABLE
SEE PROCESS CHART ON MILLING	MEDIUM	40 SQ. FT.	TABLE
SEE PROCESS CHART ON MILLING	MEDIUM	32 SQ. FT.	TABLE AND A VERTICAL SPINDLE ON A SLIDE
SEE PROCESS CHART ON MILLING	HIGH	40 SQ. FT.	KNEE, SADDLE TABLE, VERTICAL SPINDLE, AND RAM
SEE PROCESS CHART ON MILLING	HIGH	64 SQ. FT.	KNEE, SADDLE TABLE, RAM, AND SPINDLE CONTOURING
SEE PROCESS CHART ON MILLING	HIGH	64 SQ. FT.	TABLE AND VERTICAL SADDLE
SEE PROCESS CHART ON MILLING	HIGH	64 SQ. FT.	TABLE, VERTICAL, AND CROSS SADDLE

Table 12 Plastics Processing

MACHINE	APPLICATION INFORMATION		
PLASTICS PROCESSING / TYPES	1. TYPICAL APPLICATIONS	2. TYPICAL PROCESSES COMPLETED ON THIS MACHINE TYPE	3. WORK ENVELOPE OR MAXIMUM SIZE OF WORKPIECE
1. BLOW MOLDER	MAKING HOLLOW PARTS SUCH AS BOTTLES	BLOW MOLDING	100 CUBIC FEET
2. CALENDER	FORMING LONG CONTINUOUS SHEET OF PLASTICS	ROLLING OF SHEET	LIMITED ONLY BY MACHINE DESIGN
3. COMPRESSION MOLDER	FORMING PARTS FROM THERMOSET PLASTIC GRANULES	COMPRESSION MOLDING	500 OZ.
4. EXTRUDER	EXTRUSION OF ROD, CHANNEL, TUBING, ETC.	EXTRUSION	LIMITED ONLY BY MACHINE DESIGN
5. FILAMENT WINDER	MAKING OF LARGE STRONG THIN WALLED TANKS	WINDING PLASTIC FILAMENTS AROUND A FORM	15 FEET IN DIA. X 65 FEET LONG
6. HOT STAMPER	PLACING DECORATIVE STAMPS ON PLASTIC PARTS	PLACING MYLAR FILM ON PLASTIC PARTS	12 IN. X 12 IN.
7. INJECTION MOLDING	A VARIETY OF SMALL AND COMPLEX THERMOPLASTIC PARTS	INJECTION MOLDING (SIMILAR TO DIE CAST)	500 OZ.
8. LAMINATING	HELMETS, BOATS, SHEETS, AND CAR BODIES	LAMINATING OF PLASTIC, RESIN, GLASS CLOTH, ETC.	LIMITED ONLY BY MACHINE DESIGN
9. REACTION INJECTION MOLDING	CREATING OF STRUCTURAL FOAM PARTS FOR FURNITURE	MIXING OF TWO REACTIVE SUBSTANCES	LIMITED BY SIZE OF MOLD
10. ROTATIONAL MOLDING	TANKS, DRUMS, AND BUCKETS	MOLDING	6 FEET IN DIA. X 12 FEET LONG
11. TRANSFER MOLDING	INTRICATE SHAPES IN THERMOSET PLASTICS	COMPRESSION MOLDING	500 OZ.
12. VACUUM MOLDING	SIGNS, LINERS, COVERS, AND PANELS	COMPRESSION MOLDING	LIMITED ONLY BY MACHINE DESIGN

APPLICATION INFORMATION			
4. PRODUCTION RATE	5. RELATIVE EQUIPMENT COST	6. FLOOR SPACE REQUIRED	7. DISADVANTAGE
100/HOUR	HIGH	64 SQ. FT.	LIMITED TO CYLINDRICAL SHAPES OF UNIFORM WALL THICKNESS
DEPENDS ON THICKNESS OF SHEET	HIGH	200 SQ. FT.	HIGH CAPITAL EQUIPMENT COST
100/HOUR	MEDIUM	25 SQ. FT.	FLASH MUST BE REMOVED FROM THE PART
DEPENDS ON SQUARE INCHES BEING EXTRUDED	HIGH	100 SQ. FT.	HIGH CAPITAL EQUIPMENT COST
4/HOUR	MEDIUM	64 SQ. FT.	LIMITED APPLICATIONS
240/HOUR	LOW	TABLE MOUNTED	NONE
240/HOUR	HIGH	64 SQ. FT.	PARTS MUST BE TRIMMED AND DEGATED
DEPENDS UPON TYPE OF PART BEING LAMINATED	HIGH	200 SQ. FT.	REQUIRES A CURING STEP
240/HOUR	HIGH	25 SQ. FT.	TOXIC FUMES REQUIRES VENTILATION SYSTEM
100/HOUR	HIGH	25 SQ. FT.	NONE
100/HOUR	HIGH	25 SQ. FT.	FLASH MUST BE REMOVED FROM THE PART
240/HOUR	LOW	25 SQ. FT.	NONE

Table 13 Saws

MACHINE	APPLICATION INFORMATION		
SAWS / TYPES	1. TYPICAL APPLICATIONS	2. TYPICAL PROCESSES COMPLETED ON THIS MACHINE TYPE	3. WORK ENVELOPE OR MAXIMUM SIZE OF WORKPIECE
1. HACKSAW (COLUMN TYPE HORIZONTAL)	SAWING OF THICK HARD MATERIALS	SAWING ACCURATELY WITHOUT BLADE TWIST	A 12 INCH CUBE OF UNLIMITED LENGTH
2. HACKSAW (HORIZONTAL)	SAWING OF THICK HARD MATERIALS	SAWING ACCURATELY WITHOUT BLADE TWIST	A 12 INCH CUBE OF UNLIMITED LENGTH
3. HACKSAW (VERTICAL)	SAWING OF THICK HARD MATERIALS	SAWING ACCURATELY WITHOUT BLADE TWIST	A 12 INCH CUBE OF UNLIMITED LENGTH
4. HACKSAW (SCISSOR)	SAWING OF THICK HARD MATERIALS	SAWING ACCURATELY WITHOUT BLADE TWIST	A 12 INCH DIA. OF UNLIMITED LENGTH
5. BANDSAW (VERTICAL)	SAWING METAL OR WOOD	CUTOFF, CONTOUR CUTTING AND SLOTTING	CUT WIDTHS OF UP TO 12 INCHES
6. BANDSAW (HORIZONTAL)	SAWING METAL OR WOOD	CUTOFF	CUT WIDTHS OF UP TO 62 INCHES
7. BANDSAW (TILT FRAME)	SAWING METAL OR WOOD	CUTOFF	CUT WIDTHS OF UP TO 25 INCHES
8. CIRCULAR (PIVOT-ARM)	SAWING OF BARS ETC. INTO SHORT SLUGS	CUTOFF	12 INCHES THICK (28 IN. DIA. BLADE)
9. CIRCULAR (VERTICAL COLUMN)	SAWING OF BARS ETC. INTO SHORT SLUGS	CUTOFF	12 INCHES THICK (28 IN. DIA. BLADE)
10. CIRCULAR (HORIZONTAL TRAVEL)	SAWING OF BARS ETC. INTO SHORT SLUGS	CUTOFF	12 INCHES THICK (28 IN. DIA. BLADE)
11. CIRCULAR (PLATE SAWING)	SAWING OF BARS ETC. INTO SHORT SLUGS	CUTOFF	12 INCHES THICK (28 IN. DIA. BLADE)
12.			

APPLICATION INFORMATION			
4. PRODUCTION RATE	5. RELATIVE EQUIPMENT COST	6. FLOOR SPACE REQUIRED	7. DISADVANTAGE
VARIES WITH SIZE OF PART MATERIAL (I.E. 100/HOUR)	LOW	60 SQ. FT. INCLUDES STOCK	SLOWER THAN OTHER SAWING METHODS
VARIES WITH SIZE OF PART MATERIAL (I.E. 100/HOUR)	LOW	60 SQ. FT. INCLUDES STOCK	SLOWER THAN OTHER SAWING METHODS
VARIES WITH SIZE OF PART MATERIAL (I.E. 100/HOUR)	LOW	60 SQ. FT. INCLUDES STOCK	SLOWER THAN OTHER SAWING METHODS
VARIES WITH SIZE OF PART MATERIAL (I.E. 100/HOUR)	LOW	60 SQ. FT. INCLUDES STOCK	SLOWER THAN OTHER SAWING METHODS
VARIES WITH SIZE OF PART MATERIAL (I.E. 100/HOUR)	LOW	12 SQ. FT. LESS STOCK	RIGIDITY OF THE BAND IS DECREASED ON LARGE PARTS
VARIES WITH SIZE OF PART MATERIAL (I.E. 100/HOUR)	LOW	100 SQ. FT. INCLUDES STOCK	RIGIDITY OF THE BAND IS DECREASED ON LARGE PARTS
VARIES WITH SIZE OF PART MATERIAL (I.E. 100/HOUR)	LOW	60 SQ. FT. INCLUDES STOCK	RIGIDITY OF THE BAND IS DECREASED ON LARGE PARTS
VARIES WITH SIZE OF PART MATERIAL (I.E. 100/HOUR)	MEDIUM	100 SQ. FT. INCLUDES STOCK	WIDE BLADE RESULTS IN MORE MATERIAL WASTE
VARIES WITH SIZE OF PART MATERIAL (I.E. 100/HOUR)	MEDIUM	100 SQ. FT. INCLUDES STOCK	WIDE BLADE RESULTS IN MORE MATERIAL WASTE
VARIES WITH SIZE OF PART MATERIAL (I.E. 100/HOUR)	MEDIUM	100 SQ. FT. INCLUDES STOCK	WIDE BLADE RESULTS IN MORE MATERIAL WASTE
VARIES WITH SIZE OF PART MATERIAL (I.E. 100/HOUR)	MEDIUM	100 SQ. FT. INCLUDES STOCK	WIDE BLADE RESULTS IN MORE MATERIAL WASTE

Table 14 Turning

MACHINE	APPLICATION INFORMATION		
TURNING TYPES	1. TYPICAL APPLICATIONS	2. TYPICAL PROCESSES COMPLETED ON THIS MACHINE TYPE	3. WORK ENVELOPE OR MAXIMUM SIZE OF WORKPIECE
1. ENGINE LATHE	TOOLMAKING	TURNING, BORING, FACING, GROOVING, THREADING, TREPANNING CHAMFERING, ETC.	18 INCH CHUCK
2. TURRET LATHE RAM TYPE	LOW VOLUME PRODUCTION LATHE WORK	TURNING, BORING, FACING, GROOVING, THREADING, TREPANNING CHAMFERING, ETC.	18 INCH CHUCK
3. TURRET LATHE SADDLE TYPE	ACCURATE LONG CUT ON LARGER PARTS	TURNING, BORING, FACING, GROOVING, THREADING, TREPANNING CHAMFERING, ETC.	18 INCH CHUCK
4. AUTOMATIC TURRET LATHE	PRODUCTION LATHE WHERE 1 OPERATOR RUNS 2 MACHINES	TURNING, BORING, FACING, GROOVING, THREADING, TREPANNING CHAMFERING, ETC.	18 INCH CHUCK
5. N.C. LATHE	PRODUCTION LATHE WORK WITH RUNS OF 1-1,500 PARTS	TURNING, BORING, FACING, GROOVING, THREADING, TREPANNING CHAMFERING, ETC.	18 INCH CHUCK
6. MULTIPLE SPINDLE BAR MACHINE	HIGH VOLUME PRODUCTION OF COMPLEX BAR TYPE PARTS	TURNING, BORING, FACING, GROOVING, THREADING, TREPANNING CHAMFERING, ETC.	3.75 INCH CHUCK
7. MULTIPLE SPINDLE CHUCKER	HIGH VOLUME PRODUCTION OF COMPLEX CHUCKING TYPE PARTS	TURNING, BORING, FACING, GROOVING, THREADING, TREPANNING CHAMFERING, ETC.	10 INCH CHUCK
8. SWISS AUTOMATIC	PRODUCTION OF LONG SMALL DIAMETER SHAFTS	TURNING, BORING, FACING, GROOVING, THREADING, TREPANNING CHAMFERING, ETC.	1.5 INCH DIA SHAFT
9. BROWN & SHARP AUTOMATIC	HIGH VOLUME PRODUCTION OF COMPLEX YET SMALL PARTS	TURNING, BORING, FACING, GROOVING, THREADING, TREPANNING CHAMFERING, ETC.	6 INCH CHUCK
10. PRECISION BORING MACHINE	HIGH VOLUME PRODUCTION OF LATHE TYPE PARTS WITH ACCURATE BORES	TURNING, BORING, FACING, GROOVING, THREADING, TREPANNING CHAMFERING, ETC.	18 INCH CHUCK
11. HEAVY HORIZONTAL BORING MACHINE	LARGE WORK THAT WILL NOT FIT ON A LATHE	TURNING, BORING, FACING, GROOVING, THREADING, TREPANNING CHAMFERING, ETC.	A 6 FOOT CUBE
12. MULTIPLE SPINDLE VERTICAL CHUCKER	LATHE WORK THAT IS TOO HEAVY TO BE MOUNTED HORIZONTALLY	TURNING, BORING, FACING, GROOVING, THREADING, TREPANNING CHAMFERING, ETC.	12 INCH CHUCK

APPLICATION INFORMATION			
4. PRODUCTION RATE	5. RELATIVE EQUIPMENT COST	6. FLOOR SPACE REQUIRED	7. DISADVANTAGE
NOT A PRODUCTION MACHINE	LOW	24 SQ. FT.	REQUIRES SKILLED LABOR
20/HOUR	LOW	30 SQ. FT.	REQUIRES CONSTANT OPERATOR INVOLVEMENT
20/HOUR	LOW	30 SQ. FT.	REQUIRES CONSTANT OPERATOR INVOLVEMENT
30/HOUR	MEDIUM	32 SQ. FT.	SET UP TIME RELATIVELY LONG
30/HOUR	MEDIUM	32 SQ. FT. PLUS CONTROLLER	NOT SUITABLE FOR EXTREMELY LONG RUNS
150/HOUR	HIGH	100 SQ. FT.	SET UP TIME RELATIVELY LONG
150/HOUR	HIGH	100 SQ. FT.	SET UP TIME RELATIVELY LONG
200/HOUR	MEDIUM	60 SQ. FT.	LIMITED APPLICATIONS (I.E. LONG BAR TYPE WORK)
500/HOUR	MEDIUM	18 SQ. FT. FOR CHUCKER 56 SQ. FT. FOR BAR MACHINE	NONE
100/HOUR	MEDIUM	UP TO 45 SQ. FT.	NONE
4/HOUR	HIGH	100 SQ. FT.	LOAD AND UNLOAD OF PARTS OFTEN REQUIRES A CRANE
150/HOUR	HIGH	50 SQ. FT.	SET UP TIME RELATIVELY LONG

APPENDIX V

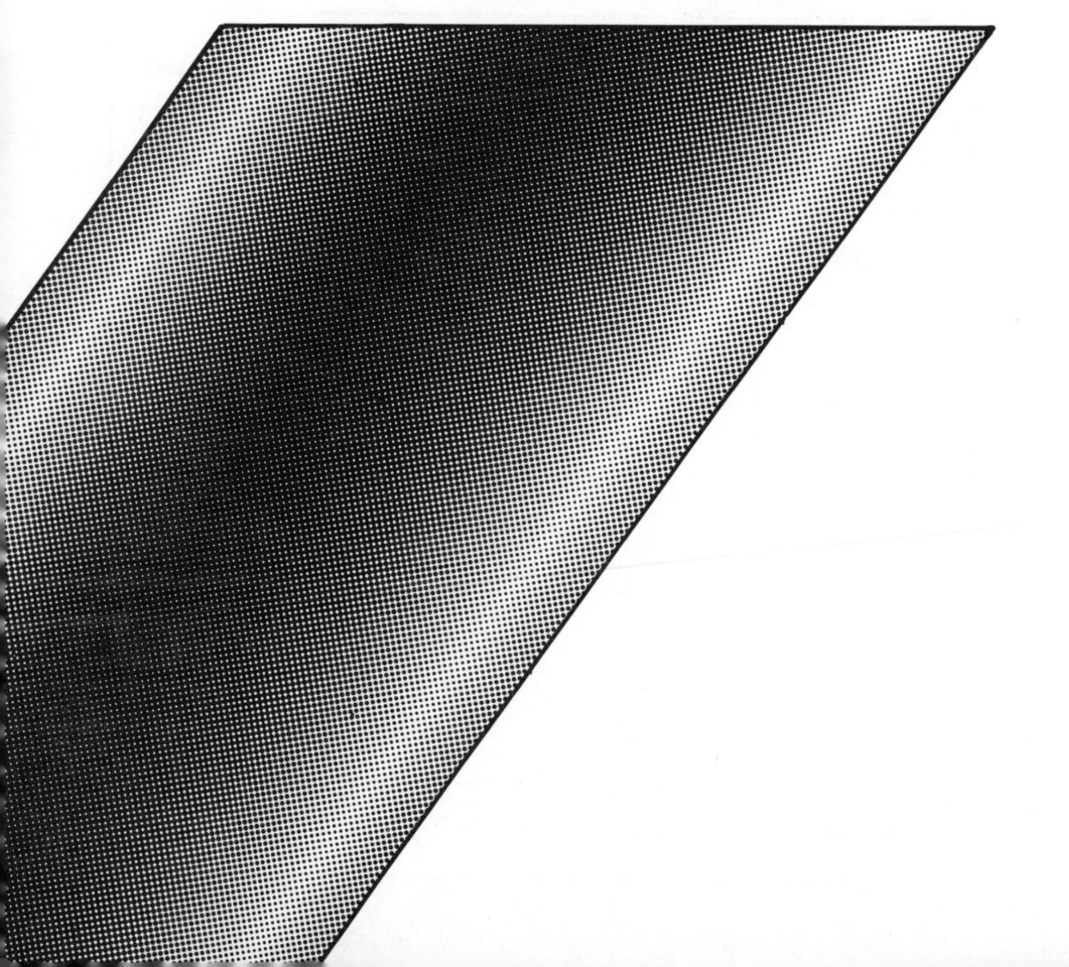

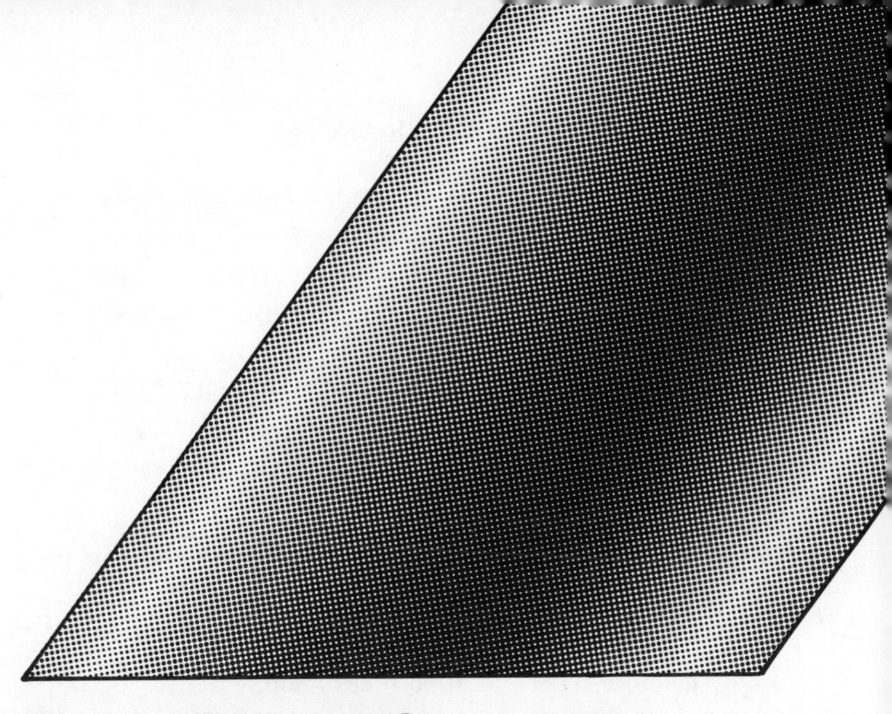

Formulas and Conversions

TRIGONOMETRY; RIGHT TRIANGLES:

$$\frac{\text{SIDE OPPOSITE}}{\text{HYPOTENUSE}} = \text{SINE}$$

$$\frac{\text{SIDE ADJACENT}}{\text{HYPOTENUSE}} = \text{COSINE}$$

$$\frac{\text{SIDE OPPOSITE}}{\text{SIDE ADJACENT}} = \text{TANGENT}$$

$$\frac{\text{SIDE ADJACENT}}{\text{SIDE OPPOSITE}} = \text{COTANGENT}$$

$$\frac{\text{HYPOTENUSE}}{\text{SIDE ADJACENT}} = \text{SECANT}$$

$$\frac{\text{HYPOTENUSE}}{\text{SIDE OPPOSITE}} = \text{COSECANT}$$

TIME AND MOTION STUDY:

A decimal minute watch divides one minute into 100 equal parts.

$$\frac{\text{1 DECIMAL MINUTE}}{\text{60 SECONDS}} = \frac{1.00}{60} = .016666 \text{ DECIMAL MINUTES}$$

Therefore (1) Sec. = .016666 Dec. Min.

SECONDS (.016666) = DECIMAL MINUTES

EFFICIENCY:

$$\frac{\text{EARNED STANDARD}}{\text{ACTUAL TIME}} \times 100 = \% \text{ EFFICIENT}$$

Where: EARNED STANDARD = STANDARD TIME × PIECES PRODUCED

ACTUAL TIME = ACTUAL TOTAL TIME REQUIRED TO PRODUCE PIECES

STATISTICS:

$$\text{AVERAGE} = \overline{X} = \frac{X_1 + X_2 + X_3 + \ldots\ldots\ldots\ldots X_n}{n}$$

$$\text{RANGE} = R = X_{\text{maximum}} - X_{\text{minimum}}$$

$$\text{VARIANCE of a SAMPLE} = S^2 = \frac{\Sigma\,(X_1 - \overline{X})^2}{n-1}$$

Where: $\overline{X}$ = MEAN OF THE SAMPLE

n = NUMBERS of ITEMS IN SAMPLE

$$\text{STANDARD DEVIATION of a SAMPLE} = S = \sqrt{S^2} = \sqrt{\frac{\Sigma (X_1 - \overline{X})^2}{n-1}}$$

SPEED; SURFACE FOOTAGE:

$$\frac{(D)(\pi)(RPM)}{12} = SFM$$

Where: D = DIAMETER in INCHES
π = Pi or 3.1416
RPM = REVOLUTIONS/MINUTE
SFM = SURFACE FEET/MINUTE or SURFACE FOOTAGE

SELECTED CONVERSIONS:

1 Acre = 43,560 Square Feet
1 Board Foot = 144 Cubic Inches
1 Centimeter = .3937 Inches
1 Foot = .3048 Meters
1 Gallon = 231 Cubic Inches
1 Gallon = 8 Pints Liquid
1 Gallon = 4 Quarts Liquid
1 Gallon Water = 8.345 Pounds
1 Gram = .03527 Ounces
1 Inch = 2.54 Centimeters
1 Kilogram = 2.205 Pounds
1 Kilometer = .6214 Miles
1 Liter = .2642 Gallons
1 Liter = 2.113 Pints Liquid
1 Meter = 1.094 Yards
1 Mile = 1.609 Kilometers
1 Millimeter = .03937 Inches
1 Ounce = 437.5 Grains
1 Pound = 256 Drams
1 Ream = 500 Sheets

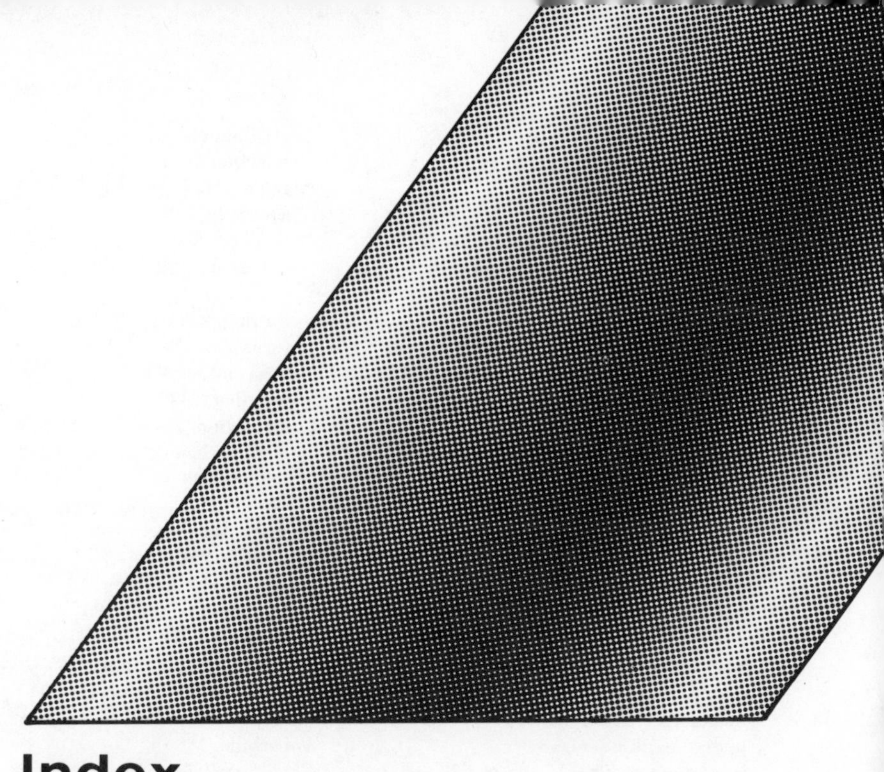

Index